Praise for *Kanban*

David's work with Kanban Systems has been a significant influence on how I approach software development and has changed the way I think about processes. Rather than viewing work in terms of stories, points, and timeboxes, I now see WIP, flow, and cadence. This book is a milestone in bringing this perspective to a wider audience, and is a must-read for anyone looking for ways of creating successful and sustainable development organizations.

—Karl Scotland
Senior Practice Consultant, EMC Consulting

Kanban is a tricky subject to write on, since everyone's implementation will be tailored to their specific workflow and bottlenecks, but David manages to provide a sound theoretical framework while maintaining a strict adherence to practicality and real-world results.

—Chris Simmons
Development Manager, Sophos

David Anderson's book *Kanban* goes beyond the introductory level of how kanban drives change, and provides clear explanation of the nuts and bolts, giving rich examples and practical tips. Kanban for knowledge work powerfully supports the emerging trend of autonomy in the workplace, one of the most exciting management developments of our time.

—Christina Skaskiw
Agile coach

Best new change methodology I have seen for software in the last ten years.

—David A. Bulkin
Vice President, Lithespeed, LLC

Kanban

Successful Evolutionary Change
for Your Technology Business

David J. Anderson

BLUE
HOLE
PRESS

Sequim, Washington

Blue Hole Press
72 Buckhorn Road
Sequim, WA 98382
www.blueholepress.com

Printed in the United States of America.

Library of Congress Cataloging-in-Publication Data
applied for; available at www.loc.gov

ISBN: 978-0-9845214-0-1 (paperback)
ISBN: 978-0-9845214-1-8 (PDF)

Cover art copyright © 2010 by Pujan Roka
Cover photo copyright © 2010 by Laurence Cohen
Cover and interior design by Vicki L. Rowland
Photos on page 12 used with permission, Thomas Blomseth

10 9 8 7 6 5 4

❖ Contents ❖

Chapter 4
From Worst to Best in Five Quarters

Chapter 5
A Continuous Improvement Culture

Chapter 9

Chapter 10

Chapter 11

PART FOUR ❖ Making Improvements

Chapter 16
Three Types of Improvement Opportunity _____ 187

Chapter 17
Bottlenecks and Non-Instant Availability _____ 195

Chapter 18
An Economic Model for Lean _____ 211

❖ *For Nicola and Natalie* ❖

❖ Foreword ❖

I always pay attention to David Anderson's work. My first contact with him was in October 2003, when he sent me a copy of his book, *Agile Management for Software Engineering: Applying Theory of Constraints for Business Results*. As its title implies, this book was heavily influenced by Eli Goldratt's Theory of Constraints (TOC). Later, in March 2005, I visited him at Microsoft; by this time he was doing impressive work with cumulative flow diagrams. Later still, in April 2007, I had a chance to observe the breakthrough kanban system that he had implemented at Corbis.

I trace this chronology to give you a sense of the relentless pace at which David's management thinking has advanced. He does not get stuck on a single idea and try to force the world to fit it. Instead, he pays careful attention to the overall problem he is trying to solve, stays open to different possible solutions, tests them in action, and reflects on why they work. You will see the results of this approach throughout this new book.

Of course, speed is most useful if it is in the correct direction; I am confident David is headed in the right direction. I am particularly excited by this latest work with kanban systems. I have always found the ideas of lean manufacturing more directly useful in product development than those of TOC. In fact, in October 2003 I wrote to David, saying, "One of the great weaknesses of TOC is its underemphasis of the importance of batch size. If your first priority is to find and reduce the constraint you are often solving the wrong problem." I still believe this is true.

In our 2005 meeting I again encouraged David to look beyond the bottleneck focus of TOC. I explained to him that the dramatic success of the Toyota Production System (TPS) had nothing to do with finding and eliminating bottlenecks. Toyota's performance gains came from using batch-size reduction and variability reduction to reduce work-in-process inventory. It was the reduction in inventory that unlocked the economic benefits, and it was WIP-constraining systems like kanban that made this possible.

By the time I visited Corbis in 2007 I saw an impressive implementation of a kanban system. I pointed out to David that he had progressed far beyond the kanban approach used by Toyota. Why did I say this? The Toyota Production System is elegantly optimized to deal with repetitive and predictable tasks: tasks with homogeneous task durations and homogeneous delay costs. Under such conditions it is correct to use approaches like first-in-first-out (FIFO) prioritization. It is also correct to block the entry of work when the WIP limit is reached. However, these approaches are not optimal when we must deal with non-repetitive, unpredictable jobs; with different delay costs; and different task durations—exactly what we must deal with in product development. We need more advanced systems, and this book is the first to describe these systems in practical detail.

I'd like to offer a few brief warnings to readers. First, if you think you already understand how kanban systems work, you are probably thinking of the kanban systems used in lean manufacturing. The ideas in this book go far beyond such simple systems that use static WIP limits, FIFO scheduling, and a single class of service. Pay careful attention to these differences.

Second, don't just think of this approach as a visual control system. The way kanban boards make WIP visible is striking, but it is only one small aspect of this approach. If you read this book carefully you will find much more going on. The real insights lie in aspects like the design of arrival and departure processes, the management of non-fungible resources, and the use of classes of service. Don't be distracted by the visual part and miss the subtleties.

Third, don't discount these methods because they appear easy to use. This ease of use is a direct result of David's insight into what produces the maximum benefit with the minimum effort. He is keenly aware of the needs of practitioners and has paid careful attention to what actually works. Simple methods create the least disruption and almost always produce the largest sustained benefits.

This is an exciting and important book that deserves careful reading. What you will get from it will depend on how seriously you read it. No other book will give you a better exposure to these advanced ideas. I hope will you enjoy it as much as I have.

Don Reinertsen,
February 7, 2010
Redondo Beach, California
Author of *The Principles of Product Development Flow*

❖ PART ONE ❖

Introduction

Chapter 1

Solving an Agile Manager's Dilemma

In 2002, I was an embattled development manager in a remote outpost of Motorola's PCS (cell phone) division based in Seattle, Washington. My department had been part of a startup company Motorola had acquired a year earlier. We developed network server software for wireless data services such as over-the-air download and over-the-air device management. These server applications were part of integrated systems that worked hand-in-hand with client code on the cell phones, as well as with other elements within the telecom carriers' networks and back-office infrastructure, such as billing. Our deadlines were set by managers without regard to engineering complexity, risk, or project size. Our code base had evolved from the original startup company, where many corners had been cut. One senior developer insisted on referring to our product as "a prototype." We were in desperate need of greater productivity and higher quality in order to meet business demands.

In my daily work, back in 2002, and through my writing efforts on my earlier book[1], two main challenges were on my mind. First, how could I protect my team from the incessant demands of the business and achieve what the Agile community now refers to as a "sustainable pace"? And second, how could I successfully scale adoption of an Agile approach across an enterprise and overcome the inevitable resistance to change?

My Search for Sustainable Pace

Back in 2002, the Agile community referred to the notion of a sustainable pace as simply "the 40-hour week."[2] The Principles Behind the Agile Manifesto[3] told us that "Agile processes promote sustainable development. The sponsors, developers, and users should be able to maintain a constant pace indefinitely." Two years earlier, my team at Sprint PCS had become used to me telling them that "large-scale software development is a marathon, not a sprint." If team members were to keep up the pace for the long haul on an 18-month project, we couldn't afford to burn them out after a month or two. The project had to be planned, budgeted, scheduled, and estimated so that team members could work reasonable hours each day and avoid tiring themselves out. The challenge for me as a manager was to achieve this goal *and* accommodate all the business demands.

In my first management job in 1991, at a five–year-old startup that made video capture boards for PCs and other, smaller computers, feedback from the CEO was that the leadership saw me as "very negative." I was always answering "No" when they asked for yet more products or features when our development capacity was already stretched to the maximum. By 2002, there was clearly a pattern: I'd spent more than ten years saying "No," pushing back against the constant, fickle demands of business owners.

In general, software engineering teams and IT departments seemed to be at the mercy of other groups who would negotiate, cajole, intimidate, and overrule even the most defensible and objectively derived plans. Even plans based on thorough analysis and backed by years of historical data were vulnerable. Most teams, which had neither a thorough analysis method nor any historical data, were powerless at the hands of others who would push them to commit to unknown (and often completely unreasonable) deliverables.

Meanwhile, the workforce had largely accepted crazy schedules and ridiculous work commitments as the norm. Software engineers are apparently not supposed to have a social or family life. If that feels abusive, it's because it is! I know too many people whose commitment to work has irreparably damaged relationships with their children and other family members. It's tough to have sympathy for the typical software development geek. In my home state of Washington, in the United States, software engineers are second only to dentists in annual income. Like Ford assembly-line workers in the second decade of the twentieth century—earning five times the average U.S. wage—no one cared about the monotony of the work or the well-being of the workers because they were so well remunerated. It's hard to imagine the emergence of organized labor in knowledge-work fields like software development, mainly because it's hard to imagine anyone addressing the root causes of the physical and psychological ills developers routinely experience. More affluent employers have been apt to add additional health-care benefits, such as massage and psychotherapy, and to provide occasional "mental health" sick days, rather than pursue the root causes of the problem. A technical writer at a well-known software firm once commented to me, "There is no stigma about being on antidepressants, everyone is doing them!" In response to this abuse, software engineers tend to acquiesce to demands, collect their fancy salaries, and suffer the consequences.

I wanted to break that mold. I wanted to find a "win-win" approach that allowed me to say "Yes," and still protect the team by facilitating a sustainable pace. I wanted to give back to my team—to give them back a social- and family life—and to improve the conditions that were causing stress-related health issues in developers as young as in their 20s. So I decided to take a stand and try to do something about these problems.

My Search for Successful Change Management

The second thing on my mind was the challenge of leading change in large organizations. I'd been a development manager at Sprint PCS and later at Motorola. In both companies, there was a real business need to develop a more agile way of working. But in both cases I had struggled to scale Agile methods across more than one or two teams.

I didn't have sufficient positional power in either case simply to impose change on a large number of teams. I was trying to influence change at the request of senior leadership, but without any positional power. I had been asked to influence peers to make changes in their teams similar to the ones I had implemented with my own team. The other teams resisted adopting techniques that were quite clearly producing better results with my team. There were probably many facets to this resistance, but the most common theme was that every other team's situation was different; my team's techniques would have to be modified and tailored to others' specific needs. By mid-2002, I had concluded that prescriptively enforcing a software development process on a team didn't work.

A process needed to be adapted for each specific situation. To do this would require active leadership on each team. This was often lacking. Even with the right leadership, I doubted that significant change could happen without a framework in place and guidance for how to tailor the process to fit different situations. Without this to guide the leader, coach, or process engineer, any adaptation was likely to be imposed subjectively, based on superstitious beliefs. This was just as likely to raise hackles and objections as imposing an inappropriate process template.

I had partly set out to address this issue with the book I was writing at the time, *Agile Management for Software Engineering.* I was asking, "Why does Agile development produce better economic outcomes than traditional approaches?" I sought to use the framework of the Theory of Constraints[4] to make the case.

While researching and writing that book, I came to realize that in some way, every situation was unique. Why should the constraining factor or bottleneck be in the same place for every team and on every project, every time? Each team is different: different sets of skills, capabilities, and experience. Each project is different: different budget, schedule, scope, and risk profile. And every organization is different: a different value chain operating in a different market, as shown in Figure 1.1. It occurred to me that this might provide a clue to the resistance to change. If proposed changes to working practices and behaviors did not have a perceived benefit, people would resist them. If those changes did not affect what the team members perceived as their constraint or limiting factor, then they would resist. Simply put, changes suggested out of context would be rejected by the workers who lived and understood the project context.

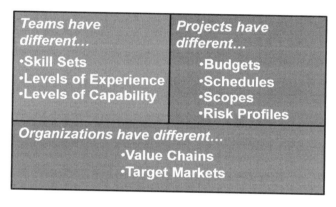

Figure 1.1   Why "one size fits all" development methodologies don't work

It seemed better to let a new process evolve by eliminating one bottleneck after another. This is the core thesis of Goldratt's Theory of Constraints. While recognizing that I had a lot to learn, I felt there was value in the material and I pressed ahead with the originally planned manuscript. I knew full well that my book did not provide advice on how to implement the ideas at scale, as it offered little or no advice on change management.

Goldratt's approach, explained in chapter 16, seeks to identify a bottleneck and then find ways to alleviate it until it no longer constrains performance. Once this happens, a new bottleneck emerges and the cycle repeats. It's an iterative approach to improving performance systematically by identifying and removing bottlenecks.

I recognized that I could synthesize this technique with some ideas from Lean. By modeling the workflow of a software development lifecycle as a value stream and then creating a tracking and visualization system to track state changes of emerging work as it "flowed" through the system, I could see bottlenecks. The ability to identify a bottleneck is the first step in the underlying model for the Theory of Constraints. Goldratt had already developed an application of the theory for flow problems, awkwardly called "Drum-Buffer-Rope." Regardless of the awkwardness of the name, I realized that a simplified Drum-Buffer-Rope solution could be implemented for software development.

Generically, Drum-Buffer-Rope is an example of a class of solutions known as pull systems. As we will see in chapter 2, a kanban system is another example of a pull system. An interesting side effect of pull systems is that they limit work-in-progress (WIP) to some agreed-upon quantity, thus preventing workers from becoming overloaded. In addition, only the workers at the bottleneck station remain fully loaded; everyone else should experience some slack time. I realized that pull systems could potentially solve both of my challenges. A pull system would enable me to implement process change incrementally, with (I hoped) significantly reduced resistance, and it would facilitate a sustainable pace. I resolved to establish a Drum-Buffer-Rope pull system at the earliest opportunity. I wanted to experiment with incremental process evolution and see whether it enabled sustainable pace and reduced resistance to change.

The opportunity arrived in the fall of 2004 at Microsoft, and it is fully documented in the case study in chapter 4.

From Drum-Buffer-Rope to Kanban

Implementing a Drum-Buffer-Rope solution at Microsoft worked well. With very little resistance, productivity more than tripled and lead times shrank by 90 percent, while predictability improved by 98 percent. By the fall of 2005 I reported the results at a conference in Barcelona[5], and again in winter of 2006. My work came to the attention of Donald Reinertsen, who made a special trip to visit me at my office in Redmond, Washington. He wanted to persuade me that I had all the pieces in place to implement a full kanban system.

Kan-ban is a Japanese word that literally means "signal card" in English. In a manufacturing environment, this card is used as a signal to tell an upstream step in a process to produce more. The workers at each step in the process are not allowed to do work unless they are signaled with a kanban from a downstream step. Although I was aware of this mechanism, I was not convinced that it was either a useful or a viable technique for application to knowledge work and specifically software engineering. I understood that a kanban system would enable a sustainable pace. However, I was not aware of its reputation as a method for driving incremental process improvement. I was unaware that Taiichi Ohno, one of the creators of the Toyota Production System had said, "The two pillars of the Toyota production system are just-in-time and automation with a human touch, or autonomation. The tool used to operate the system is kanban." In other words, kanban is fundamental to the *kaizen* ("continuous improvement") process used at Toyota. It is the mechanism that makes it work. I have come to recognize this as a complete truth through my experiences over the five years since.

Luckily, Don made a convincing argument that I should switch from Drum-Buffer-Rope implementations to a kanban system for the highly esoteric reason that a kanban system makes a more graceful recovery from an outage in the bottleneck station than Drum-Buffer-Rope system does. Understanding this idiosyncrasy, however, is not important for you to be able to read and understand this book.

Revisiting the final implemented solution at Microsoft, I realized that had we conceived of it as a kanban system from the beginning, the outcome would have been identical. It was interesting to me that two different approaches would result in the same outcome. So if the resultant process was the same, I didn't feel compelled to think of it specifically as a Drum-Buffer-Rope implementation.

I developed a preference for the term "kanban" over Drum-Buffer-Rope. Kanban is used in Lean manufacturing (or the Toyota Production System). This body of knowledge has much wider adoption and acceptance than the Theory of Constraints. Kanban, while Japanese, is less metaphorical than Drum-Buffer-Rope. Kanban was easier to say, easier to explain, and, as it turned out, easier to teach and implement, so it stuck.

Emergence of the Kanban Method

In September 2006, I moved from Microsoft to take over the software engineering department at Corbis, a privately held stock photography and intellectual property rights business based in downtown Seattle. Encouraged by the results from Microsoft, I decided to implement a kanban pull system at Corbis. Again, the results were encouraging and led to the development of most of the ideas presented in this book. It is this expanded set of ideas—workflow visualization, work item types, cadence, classes of service, specific management reporting, and operations reviews—that defines the Kanban Method.

In the remainder of this book I describe Kanban (capital K) as the evolutionary change method that utilizes a kanban (small k) pull system, visualization, and other tools to catalyze the introduction of Lean ideas into technology development and IT operations. The process is evolutionary and incremental. Kanban enables you to achieve context-specific process optimization with minimal resistance to change while maintaining a sustainable pace for the workers involved.

Kanban's Community Adoption

In May 2007, Rick Garber and I presented the early results from Corbis at the Lean New Product Development conference in Chicago to an audience of around 55 people. Later that summer, at the Agile 2007 conference in Washington, D.C., I held an open-space session to discuss kanban systems; about 25 people attended. Two days later, one of the attendees, Arlo Belshee, gave a lightning talk in which he shared his Naked Planning[6] technique. It appeared that others had been implementing pull systems, too. A Yahoo! discussion group was formed and quickly grew to 100 members. At the time of this writing it has over 1,000 members. Several of the attendees from the open space session committed to trying Kanban in their workplace, often with teams who had struggled with Scrum. The most notable of those early adopters were Karl Scotland, Aaron Sanders, and Joe Arnold, all from Yahoo!, who quickly took Kanban to more than ten teams on three continents. Another notable attendee at the open space was Kenji Hiranabe, who had been developing kanban solutions in Japan. Soon afterward he wrote two articles on the topic for InfoQ[7,8] that attracted a lot of interest and attention. In the fall of 2007 Sanjiv Augustine, author of *Managing Agile Projects*[9] and a founder of the Agile Project Leadership Network (APLN) visited Corbis in Seattle and described our kanban system as the "first new agile method I've seen in five years."

The following year, at Agile 2008, in Toronto, there were six presentations about the use of kanban solutions in different settings. One of them, from Joshua Kerievsky, of Industrial Logic, an Extreme Programming consulting and training firm, showed how he had evolved similar ideas to adapt and improve Extreme Programming to his business context. That year, the Agile Alliance presented the Gordon Pask Award to Arlo Belshee and Kenji Hiranabe for their contributions to the Agile community. Both had made either a visible contribution to the emergence of Kanban or produced and communicated remarkably similar ideas on the subject.

The Value of Kanban is Counter-Intuitive

In many ways, knowledge work is the antithesis of a repetitive production activity. Software development is most certainly not like manufacturing. The domains exhibit wildly different attributes. Manufacturing has low variability, while much of software development is highly variable and seeks to exploit variability through novelty in design in order to drive profit. Software is by nature "soft" and often can be changed easily and cheaply, while manufacturing tends to center on "hard" things that are difficult to change. It's natural to be skeptical about the value of kanban systems in software development and other IT work. Much of what we, as a community, have learned about Kanban over the last few years is counter-intuitive. No one predicted the effect on corporate culture or the improved cross-functional collaboration that occurred at Corbis (which is described in chapter 5). In these pages, I hope to show you that "Kanban can!" I hope to convince you that by employing Kanban's simple rules you can improve productivity, predictability, and customer satisfaction, as well as reduce delivery times. And with all that, the culture of your organization will change as the increase in collaborative work helps establish better, more functional working relationships across your organization.

Takeaways

❖ Kanban systems are from a family of approaches known as *pull* systems.

❖ Eliyahu Goldratt's Drum-Buffer-Rope application of the Theory of Constraints is an alternative implementation of a pull system.

❖ The motivation for pursuing a pull-system approach was two-fold: to find a systematic way to achieve a sustainable pace of work, and to find an approach to introducing process changes that would meet with minimal resistance.

❖ Kanban is the mechanism that underpins the Toyota Production System and its *kaizen* approach to continuous improvement.

❖ The first virtual kanban system for software engineering was implemented at Microsoft beginning in 2004.

❖ Results from early Kanban implementations were encouraging with regard to achieving sustainable pace, minimizing resistance to change through an incremental evolutionary approach, and producing significant economic benefits.

❖ The Kanban Method as an approach to change started to grow in community adoption after the Agile 2007 conference in Washington, D.C., in August 2007.

❖ Throughout this text, "kanban" (small "k") refers to signal cards, and "kanban system" (small "k") refers to a pull system implemented with (virtual) signal cards.

❖ Kanban (capital "K") is used to refer to the methodology of evolutionary, incremental process improvement that emerged at Corbis from 2006 through 2008 and has continued to evolve in the wider Lean software development community in the years since.

❖ **Chapter 2** ❖

What Is the
Kanban Method?

In the spring of 2005, I had the good fortune to take a vacation in Tokyo, Japan, in early April, during cherry-blossom season. To enjoy this spectacle I made my second-ever visit to the East Gardens at the Imperial Palace in downtown Tokyo. It was here that I had a revelation—kanban wasn't only for manufacturing.

On Saturday, April 9, 2005, I entered the park via the north entrance, crossing the bridge over the moat close to the Takebashi subway station. Many Tokyoites were taking the opportunity on a sunny Saturday morning to enjoy the tranquility of the park and the beauty of the sakura (cherry blossom).

The practice of having a picnic under the cherry trees while the blossoms fall around you is known as hanami (flower party). It's an ancient tradition in Japan—a chance to reflect on the beauty, fragility, and shortness of life. The brief life of the cherry blossom is a metaphor for our own life, and our short, beautiful, and fragile existence amid the vastness of the universe.

The cherry blossoms provided contrast against the gray buildings of downtown Tokyo, its hustle and bustle, throbbing crowds of busy people, and traffic noise. The gardens were an oasis at the heart of the concrete jungle. As I crossed the bridge with my family, an elderly Japanese gentleman with a satchel over his shoulder approached us. Reaching into his bag, he produced a handful of plastic cards. He offered one to each of us, pausing briefly to decide whether my three-month-old daughter strapped to my chest required a card. He decided she did and handed me two cards. He said nothing, and, as my Japanese is limited, I offered no conversation. We walked on into the gardens to look for a spot to enjoy our family picnic.

Two hours later, after a pleasant morning in the sunshine, we packed up our picnic things and headed toward the exit at the East Gate at Otemachi. As we approached the exit, we joined a line of people in front of a small kiosk. As the line shuffled forward I saw people returning their plastic entrance cards. I fished around in my pocket and retrieved the cards I'd been given. Approaching the kiosk I saw a neatly uniformed

Japanese lady inside. Between us was a glass screen with a semi-circular hole cut out of it at counter level, very similar to an admission booth at a cinema or amusement park. I slid my plastic cards across the countertop through the hole in the glass. The lady took them in her white-gloved hands and stacked them in a rack with others. She bowed her head slightly and thanked me with a smile. No money changed hands. No explanation was given for why I'd been carrying around two white plastic admission cards since entering the park two hours earlier.

What was going on with these admission tickets? Why bother to issue a ticket if no fee was charged? My first inclination was that it must be a security scheme. By counting all the returned cards, the authorities could ensure that no stray visitors had remained inside the grounds when they closed the park in the late afternoon. Upon quick reflection I realized that would be a very poor security system. Who was to say that I'd been issued two cards rather than just one? Did my three-month-old count as luggage or a visitor? There seemed to be too much variability in the system. Too many opportunities for errors! If it were a security scheme then surely it would fail and produce false positives every day. (As a brief aside, such as system cannot readily produce false negatives, as it would require the manufacture of additional admission tickets. This is a useful common attribute of kanban systems.) Meanwhile, the troops would be out scurrying around the bushes every evening looking for lost tourists. No, it had to be something else. I realized then that the Imperial Palace Gardens was using a kanban system!

This hugely enlightening epiphany allowed me to think beyond manufacturing with respect to kanban systems. It seemed likely that kanban tokens were useful in all sorts of management situations.

Photos this page courtesy of Thomas Blomseth

What is a Kanban System?

A number of kanban (or cards) equivalent to the (agreed) capacity of a system are placed in circulation. One card attaches to one piece of work. Each card acts as a signaling mechanism. A new piece of work can be started only when a card is available. This free card is attached to a piece of work and follows it as it flows through the system. When there are no more free cards, no additional work can be started. Any new work must wait in a queue until a card becomes available. When some work is completed, its card is detached and recycled. With a card now free, a new piece of work in the queuing can be started.

This mechanism is known as a pull system because new work is pulled into the system when there is capacity to handle it, rather than being pushed into the system based on demand. A pull system cannot be overloaded if the capacity, as determined by the number of signal cards in circulation, has been set appropriately.

In the Imperial Palace Gardens, the gardens themselves are the system: The visitors are the work-in-progress, and the capacity is limited by the number of admission cards in circulation. Newly arriving visitors gain admission only when there are available tickets to hand out. On a normal day this is never an issue. However, on busy days, such as a holiday or a Saturday during cherry-blossom season, the park is popular. When all the admission tickets are given out, new visitors must queue outside across the bridge and wait for cards to be recycled from visitors as they leave. The kanban system provides a simple, cheap, and easily implemented method for controlling the size of the crowd by limiting the number of people inside the park. This allows the park wardens to maintain the gardens in good condition and avoid damage caused by too much foot traffic and overcrowding.

Kanban Applied in Software Development

In software development, we are using a virtual kanban system to limit work-in-progress. Although "kanban" means "signal card," and there are cards used in most Kanban implementations in software development, these cards do not actually function as signals to pull more work. Instead, they represent work items. Hence the term "virtual," because there is no physical signal card. The signal to pull new work is inferred from the visual quantity of work-in-progress subtracted from some indicator of the limit (or capacity). Some practitioners have implemented physical kanban using techniques such as sticky clips or physical slots on a board. More often the signal is generated from a software work-tracking system. Chapter 6 provides examples of the mechanics of kanban systems as they apply to IT work.

Card walls have become a popular visual control mechanism in Agile software development, as shown in Figure 2.1. Using either a cork notice board with index cards pinned to a board, or a whiteboard with sticky notes to track work-in-progress (WIP) has become commonplace. It's worth observing at this early stage that despite some commentary in the community to the contrary, these card walls are not inherently kanban systems. They are merely visual control systems. They allow teams to visually observe work-in-progress and to self-organize, assign their own tasks, and move work from a backlog to complete without direction from a project or line manager. However, if there is no explicit limit to work-in-progress and no signaling to pull new work through the system, it is not a kanban system. This is more fully explained in chapter 7.

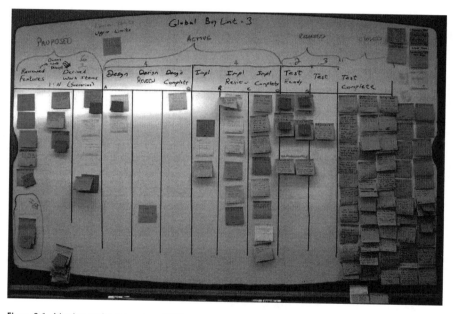

Figure 2.1 A kanban card wall (courtesy of SEP)

Why Use a Kanban System?

As should become evident in subsequent chapters, we use a kanban system to limit a team's work-in-progress to a set capacity and to balance the demand on the team against the throughput of their delivered work. By doing this we can achieve a sustainable pace of development so that all individuals can achieve a work versus personal life balance. As you will see, Kanban also quickly flushes out issues that impair performance, and it challenges a team to focus on resolving those issues in order to maintain a steady flow of work. By providing visibility onto quality and process problems, it makes obvious the impact of defects, bottlenecks, variability, and economic costs on flow and throughput.

The simple act of limiting work-in-progress with kanban encourages higher quality and greater performance. The combination of improved flow and better quality helps to shorten lead times and improve predictability and due-date performance. By establishing a regular release cadence and delivering against it consistently, Kanban helps to build trust with customers and trust along the value stream with other departments, suppliers, and dependent downstream partners.

By doing all of this, Kanban contributes to the cultural evolution of organizations. By exposing problems, focusing an organization on resolving them, and eliminating their effects in future, Kanban facilitates the emergence of a highly collaborative, high-trust, highly empowered, continuously improving organization.

Kanban has been shown to improve customer satisfaction through regular, dependable, high-quality releases of valuable software. It also has been shown to improve productivity, quality, and lead times. In addition, there is evidence that kanban is a pivotal catalyst for the emergence of a more agile organization through evolutionary cultural change.

The remainder of this book is dedicated to helping you understand how to use kanban systems in software development and to teaching you how to implement such a system to achieve these benefits with your team.

Kanban as a Complex Adaptive System for Lean

The Kanban Method introduces a complex adaptive system that is intended to catalyze a Lean outcome within an organization. Complex adaptive systems have initial conditions and simple rules that are required in order to seed complex, adaptive, emergent behavior. Kanban uses five core properties to create an emergent set of Lean behaviors in organizations. These properties have been present in every successful Kanban implementation, including the one at Microsoft described in chapter 4. The five properties are:

1. Visualize Workflow

2. Limit Work-in-Progress

3. Measure and Manage Flow

4. Make Process Policies Explicit

5. Use Models* to Recognize Improvement Opportunities

---

*   Common models in use with Kanban include the Theory of Constraints, Systems Thinking, an understanding of variability through the teachings of W. Edwards Deming, and the concept of *muda* (waste) from the Toyota Production System. The models used with Kanban are continually evolving, and ideas from other fields, such as sociology, psychology, and risk management are appearing in some implementations.

Emergent Behavior with Kanban

There is a growing list of emergent behaviors we have come to expect with Kanban implementations. Some or all of these have appeared in most of the recent implementations; all of them emerged at Corbis during 2007. We expect this list to grow as we learn more about the effects of Kanban on organizations.

1. Process uniquely tailored to each project/value stream
2. Decoupled Cadences (or "Iterationless" Development)
3. Work Scheduled by (opportunity) Cost of Delay
4. Value Optimized with Classes of Service
5. Risk Managed with Capacity Allocation
6. Tolerance for Process Experimentation
7. Quantitative Management
8. Viral Spread (of Kanban) across the Organization
9. Small teams merged for more liquid labor pools

Kanban as a Permission Giver

Kanban is not a software development lifecycle methodology or an approach to project management. It requires that some process is already in place so that Kanban can be applied to incrementally change the underlying process.

This evolutionary approach, promoting incremental change, has been controversial in the Agile software development community. It is controversial because it suggests that teams should not adopt a defined method or process template. An industry of services and tools has developed around a small set of practices defined in two popular agile development methods. Now with Kanban, individuals and teams might be empowered to evolve their own unique process solutions that obviate the need for such services and require a new set of tools. Indeed, Kanban has encouraged a new wave of insurgent tool vendors eager to displace existing agile project-management tools with something more visual and programmable that can be readily tailored to a specific workflow.

In the early days of Agile software development, the leaders in the community often didn't understand why their methods worked. We talked about "ecosystems"[10] and advised that implementers follow all of the practices or the solution was likely to fail. In recent years there has been a negative trend that extends this thinking. A few businesses have published Agile Maturity Models that seek to assess practice adoption. The Scrum community has a practice-based test often referred to as the "Nokia Test[11]."

These practice-based assessments are designed to drive conformity and deny the need for context-based adaptation. Kanban is giving the market permission to ignore these practice-based appraisal schemes. It's actively encouraging diversity.

In 2007, several people visited my offices at Corbis to see Kanban in action. The common question from any visitor associated with the Agile software development community could be paraphrased as, "David, we have been around your building and we've seen seven kanban boards. Each one is different! Each team is following a different process! How can you possibly cope with this complexity?" My answer was always a dismissive "Of course! Each team's situation is different. They evolve their process to fit their context." But I knew that these processes were derived from the same principles and that because team members understood those basic principles, they were therefore capable of adapting when reassigned from one team to another.

As more people have tried Kanban, they have realized that it helps address problems they have encountered with change management in their organizations. Kanban has enabled their team, project, or organization to exhibit better agility. We've come to recognize that Kanban is giving permission in the market to create a tailored process optimized to a specific context. Kanban is giving people permission to think for themselves. It is giving people permission to be different: different from the team across the floor, on the next floor, in the next building, and at a neighboring firm. It's giving people permission to deviate from the textbook. Best of all, Kanban is providing the tools that enable us to explain (and justify) why being different is better and why a choice to be different is the right choice in that context. To emphasize this choice, I designed a T-shirt for the Limited WIP Society, inspired by Shepard Fairey's Obama campaign poster, and featuring the face of Taiichi Ohno, the creator of the kanban system at Toyota. The slogan "Yes We Kanban" is designed to emphasize that you have permission. You have permission to try Kanban. You have permission to modify your process. You have permission to be different. Your situation is unique and you deserve to develop a unique process definition tailored and optimized to your domain, your value stream, the risks that you manage, the skills of your team, and the demands of your customers.

Takeaways

- ❖ Kanban systems can be used in any situation in which there is a desire to limit a quantity of things inside a system.

- ❖ The Imperial Palace Gardens in Tokyo uses a kanban system to control the size of the crowd inside the park.

- ❖ The quantity of "kanban" signal cards in circulation limits work-in-progress.

- ❖ New work is pulled into the process by a returning signal card upon completion of a current work order or task.

- ❖ In IT work we are (generally) using a virtual kanban system as no actual physical card is passed around to define the limit to work-in-progress.

- ❖ Card walls common in agile software development are not kanban systems.

- ❖ Kanban systems create a positive tension in the workplace that forces discussion of problems.

- ❖ The Kanban Method (or capital K "Kanban") utilizes a kanban system as a catalyst of change.

- ❖ Kanban requires that process policies are defined explicitly.

- ❖ Kanban uses tools from various fields of knowledge to encourage analysis of problems and discovery of solutions.

- ❖ Kanban enables incremental process improvement through repeated discovery of issues affecting process performance.

- ❖ A contemporary definition of the Kanban Method can be found online at the Limited WIP Society community web site, http://www.limitedwipsociety.org/.

- ❖ Kanban is acting as a permission giver in the software development profession, encouraging teams to devise context-specific process solutions rather than dogmatically following a software development lifecycle process definition or template.

❖PART TWO❖

Benefits of Kanban

❖ **Chapter 3** ❖

A Recipe
for Success

Over the past decade, I've been challenged to answer the following question: As a manager, what actions should you take when you inherit an existing team, especially one that is not working in an agile fashion, has a broad spread of ability, and is, perhaps, completely dysfunctional? Typically, I've been put in management positions as a change agent. So I've been challenged to create positive change and make progress quickly, within two or three months.

As a manager in larger organizations, I've never been able to hire my own team. I've always been asked to adopt an existing team and, with minimal personnel changes, create a revolution in the organization's performance. I think this situation is much more common than one in which you get to hire a whole new team.

I gradually evolved an approach to managing this change. It's based on experience, which includes learning from failure. The failures resulted from trying to use positional power to impose a process and a workflow. Management edict tended to fail. When I asked teams to change their behavior and use an agile method such as Feature Driven Development, I met resistance. I countered by suggesting that no one should fear, for I would provide them with the training and coaching required. I got acquiescence at best, not true, deep institutionalization of the change. Asking people to change their behavior creates fear and lowers self-esteem, as it communicates that existing skills are clearly no longer valued.

I developed what I've come to call my Recipe for Success to address these issues. The Recipe for Success presents guidelines for a new manager adopting an existing team. Following the recipe enables quick improvement with low levels of team resistance. I want to acknowledge here the direct influence of Donald Reinertsen, who gave me the first two and the last steps in the recipe, and the indirect influence of Eli Goldratt, whose writings on the Theory of Constraints and its Five Focusing Steps greatly influenced steps four and five.

The six steps in the recipe are

- Focus on Quality

- Reduce Work-in-Progress

- Deliver Often

- Balance Demand against Throughput

- Prioritize

- Attack Sources of Variability to Improve Predictability

Implementing the Recipe

The recipe is delivered in order of execution for a technical function manager. Focus on Quality is first, as it is under the sole control and influence of a manager such as a development or test manager, or the manager's supervisor, with a title like Director of Engineering. Working down the list, there is gradually less control and more collaboration required with other downstream and upstream groups until the Prioritize step. Prioritization is rightly the job of the business sector, not the technology organization, and so should not be within a technical manager's remit. Unfortunately, it is commonplace for business management to abdicate that responsibility and leave a technical manager to prioritize the work—and then blame that manager for making poor choices. Attack Sources of Variability to Improve Predictability is last on the list because some types of variability require behavioral changes in order to reduce them. Asking people to change behavior is difficult! So attacking variability is better left until a climate change results from some success with the earlier steps. As we will see in chapter 4, it is sometimes necessary to address sources of variability in order to enable some of those earlier steps. The trick is to pick sources of variability that require little behavioral change and can be readily accepted.

Focusing on quality is easiest because it is a technical discipline that can be directed by the function manager. The other steps are more of a challenge because they depend on agreement and collaboration from other teams. They require skills in articulation, negotiation, psychology, sociology, and emotional intelligence. Building consensus around the need to Balance Demand against Throughput is crucial. Solving issues with dysfunction between roles and the responsibilities of team members requires even greater diplomatic and negotiation skills. It makes sense, then, to go after things that are directly under your control and that you know will have a positive effect on both your team's and your business's performance.

Developing an increased level of trust with other teams can enable the harder things. Building and demonstrating high quality code with few defects improves trust. Releasing high-quality code with regularity builds yet more trust. As the level of trust increases, the manager gains more political capital. This enables a move to the next step in the recipe. Ultimately your team will gain sufficient respect that you are able to influence product owners, your marketing team, and business sponsors to change their behavior and collaborate to prioritize the most valuable work for development.

Attacking sources of variability to improve predictability is hard. It should not be undertaken until a team is already performing at a more mature and greatly improved level. The first four steps in the recipe will have a significant impact. They will deliver success for a new manager. However, to truly create a culture of innovation and continuous improvement, it will be necessary to attack the sources of variability in the process. So the final step in the recipe is for extra credit: It's the step that separates the truly great technical leaders from the merely competent managers.

Focus on Quality

The Agile Manifesto doesn't say anything about quality, although the Principles Behind the Manifesto[12] do talk about craftsmanship, and there is an implied focus on quality. So if quality doesn't appear in the Manifesto, why is it first in my recipe for success? Put simply, excessive defects are the biggest waste in software development. The numbers on this are staggering and show several orders of magnitude variation. Capers Jones[13] reports that in 2000, during the dot-com bubble, he measured software quality for North American teams that ranged from 6 defects per function point down to less than 3 per 100 function points, a range of 200 to 1. The midpoint is approximately 1 defect per 0.6 to 1.0 function points. This implies that it is common for teams to spend more than 90 percent of their effort fixing defects. As direct evidence, late in 2007, Aaron Sanders, an early proponent of Kanban, reported, in the Kanbandev Yahoo! group, that a team he was working with was spending 90 percent of available capacity on defect fixes.

Encouraging high initial quality will have a big impact on the productivity and throughput of teams with high defect rates. A two- to four-times throughput improvement is reasonable. With truly bad teams, a ten-times improvement may be possible by focusing on quality.

Improving software quality is a well-understood problem.

Both agile development and traditional approaches to quality have merit. They should be used in combination. Professional testers should do testing. Using testers finds defects and prevents them from escaping into production code. Asking developers to write unit tests and automate them to provide automated regression testing also has a dramatic effect. There seems to be a psychological advantage in asking developers to write the tests first. So-called Test Driven Development (TDD) does seem to provide the advantage that test coverage is more complete. It is worth pointing out that well-disciplined teams I've managed who wrote tests after functional coding demonstrated industry-leading quality. Nevertheless, it appears evident that, for the average team, insisting on writing tests first, before functional coding, improves quality.

Code inspections improve quality. Whether it is pair programming, peer review, code walkthroughs, or full Fagan inspections, code inspections work. They help to improve both external quality and internal code quality. Code inspections are best done often and in small batches. I encourage teams to inspect code every day for at least 30 minutes.

Collaborative analysis and design improve quality. When teams are asked to work together to analyze problems and design solutions, the quality is higher. I encourage teams to hold collaborative team analysis and design modeling sessions. Design modeling should be done in small batches every day. Scott Ambler has called this Agile Modeling[14].

The use of design patterns improves quality. Design patterns capture known solutions to known problems. Design patterns ensure that more information is available earlier in the lifecycle and that design defects are eliminated.

The use of modern development tools improves quality. Many modern tools include functions to perform static and dynamic code analysis. This should be switched on and tuned for each project. These analysis tools can prevent programmers from making elementary mistakes, such as introducing well-understood problems like security flaws.

More exotic modern development tools such as Software Product Lines (or Software Factories) and Domain Specific Languages reduce defects. Software Factories can be used to encapsulate design patterns as code fragments, thereby reducing the defect-insertion potential from entering code. They also can be used to automate away repetitive coding tasks; again, reducing the defect-insertion potential of entering code. The use of software factories also reduces the demands on code inspections, as factory code doesn't need to be re-inspected. It has a known quality.

Some of these latter suggestions really fall into the category of reducing variability in the process. The use of software factories and perhaps even design patterns is asking developers to change their behavior. The big bang for the buck comes from using professional testers; writing tests first, automating regression testing, and code inspections. And one more thing…

Reducing the quantity of design-in-progress boosts software quality.

Reduce Work-in-Progress and Deliver Often

In 2004 I was working with two teams at Motorola. Both teams were developing network server-side code for cell phone applications. One team was working on the over-the-air (OTA) download server for ringtones, games, and other applications and data. The other team was developing the server for over-the-air device management (OTA DM). Both teams were using the Feature Driven Development (FDD) methodology. Both teams were approximately the same size—about eight developers, one architect, up to five testers, and a project manager. Working together with the marketing people, the teams were responsible for their own analysis and design. In addition, there were user-experience design and user-documentation (technical writing) teams that provided services to both project teams.

WIP, Lead Time, and Defects

Figure 3.1 shows a cumulative flow diagram for the project work by the OTA download team. A cumulative flow diagram is an area graph that depicts the quantity of work in a given state. The states shown on this chart are: "inventory," which means backlog or queue yet to be started; "started," which implies the requirements for a feature have been explained to the developers; "designed," which means specifically that a UML sequence diagram has been developed for the feature; "coded," which means the methods on the sequence diagram have been implemented; and "complete," which means that all unit tests are passing, that the code has been peer reviewed, and that the team-lead developers have accepted the code and promoted it for testing.

The first line on the graph shows the number of features in scope for the project. The scope arrived in two batches from the business owners. The second line shows the number of features started. The third line shows the number designed. The fourth shows the number developed, and the fifth line shows the number completed and ready for testing.

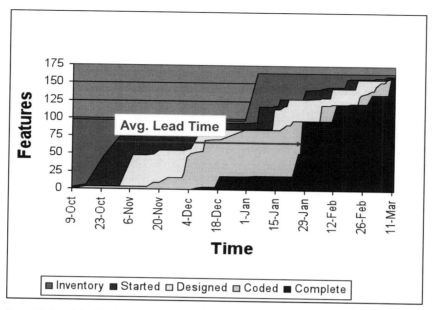

Figure 3.1 Cumulative Flow Diagram (CFS) OTA Download team fall 2003, winter 2004

The height between the second line and fifth lines on any given day shows the quantity of work-in-progress, while the horizontal distance between the second and the fifth lines shows the average lead time from starting a feature on that day until it was finished. It is important to note that the horizontal distance is an average lead time, not a specific lead time for a specific feature. The cumulative flow diagram does not track specific features. The fifty-fifth feature started might be the thirtieth feature completed. There is no association between a line on the y-axis and a specific feature in the backlog.

The OTA download server team lacked the discipline or perhaps lacked buy-in to use the FDD method. They did not work collaboratively, as FDD demands. They handed out large batches of features to individual developers. Typically, they had ten features per developer in progress at any given time. The OTA DM team followed the method to the textbook definition. They were working well collaboratively. They were developing unit tests for 100 percent of the functionality. And most importantly, they were working on small batches of features at a time, typically five to ten features in progress for the whole team at any given time. As a benchmark, a feature in FDD appears to represent about 1.6 to 2.0 function points of code.

The OTA download server team had an average lead time of around three months per unit from starting a feature to completing it for handoff from the team in Seattle, Washington, to integration test in Champaign, Illinois, as shown in Figure 3.1. The

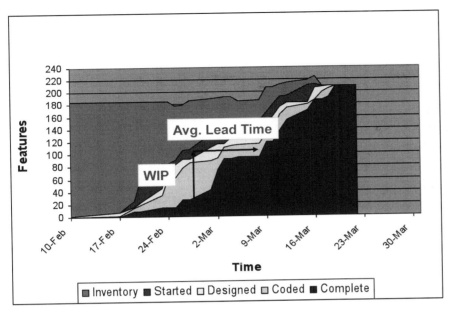

Figure 3.2 Cumulative Flow Diagram from OTA DM team winter 2004

OTA DM team had an average lead time per unit in the range of five to ten days, illustrated in Figure 3.2. The difference in initial quality, measured as escaped defects leaking into the system or integration test, was greater than 30-fold between the two teams. The OTA DM team produced industry-leading initial quality of two or three defects per 100 features, while the OTA download server team produced industry-average performance of around two defects per feature.

We can see from examining the charts that the quantity of work-in-progress is directly related to lead time. Figure 3.2 clearly shows that average lead time falls as the quantity of work-in-progress falls. At peak, the average lead time is twelve days. Later in the project, with less work-in-progress, the average lead time falls to as little as four days.

There is causation between quantity of work-in-progress and average lead time, and the relationship is linear. In the manufacturing industry, this relationship is known as Little's Law. The evidence from these two teams at Motorola suggests that there is a correlation between increased lead time and poorer quality. Longer lead times seem to be associated with significantly poorer quality. In fact, an approximately six-and-a-half times increase in average lead time resulted in a greater than 30-fold increase in initial defects. Longer average lead times result from greater amounts of work-in-progress. Hence, the management leverage point for improving quality is to reduce the quantity of work-in-progress. Since uncovering this evidence I have managed work-in-progress as a means to control quality and have become convinced of the relationship between the quantity of WIP and initial quality. However, at the time of writing there is no scientific evidence to back up this empirically observed result.

Reducing work-in-progress, or shortening the length of an iteration, will have a significant impact on initial quality. It appears that the relationship between quantity of work-in-progress and initial quality is non-linear; that is, defects will rise disproportionately to increases in quantity of WIP. Therefore, it makes sense that two-week iterations are better than four-week iterations and that one-week iterations are better still. Shorter iterations will drive higher quality.

Following the logic of the evidence presented, it makes even more sense simply to limit WIP using a kanban system. If we know that managing WIP will improve quality, why not introduce explicit policy to limit WIP, thus freeing managers to focus on other activities?

Because of the close interaction between work-in-progress and quality, it follows that step 2 of the recipe should be implemented along with or soon after step 1.

Who's better?

I intervened with the OTA download team during the week of Christmas 2003 and suggested to the team lead that there was too much work-in-progress, that the lead time was too long, and that very few things were actually being completed. I shared my concern that poor quality would result from this. He took my advice to heart and in January 2004 made some changes to the way the team worked. The result is reduced WIP in 2004 and a demonstrably shorter lead time. This change, however, came too late to prevent the team from creating a large number of defects.

While the diagram suggests that the project was completed around mid-March 2004, the team continued working on the software until mid-July that year. Half of the OTA DM team was pulled off their project to help fix the defects. In July 2004 the general manager of the business unit declared the product complete despite continued concerns about quality. The product was handed off to the field engineering team. During deployment, as many as 50 percent of customers cancelled delivery due to quality concerns. While the field engineering team maintained good personal relationships with this product's engineering team, they had a low level of professional respect and lacked trust in the team's ability. Their opinion was that the product was of poor quality and the team was incapable of delivering anything better.

Ironically, if you had walked into that building in Seattle's SODO neighborhood back then and asked the developers, "Who's the smartest guy around here?" They would have pointed to someone from the OTA download team. If you'd have asked them,

"Who has the most experience?" again, the same answer. If you'd have looked at the résumés, you would have seen that the average experience on that OTA download team exceeded that of the OTA DM team by three years. On paper, everything suggested that the OTA download team was better. And to this day, some of these individuals believe that they were better despite all the quantitative evidence otherwise.

I know from my managerial and coaching experience with this team that some of the OTA DM team members suffered from low professional self-esteem and worried that they weren't as talented as some of the truly smart folks on the other team. However, the OTA DM's productivity was five-and-a-half times greater with more than thirty times better initial quality. The right process, with good discipline, strong management, and good leadership all made a difference. What this example really demonstrates is that you don't need the best people to produce world-class results. A belief of some in the Agile community, what I refer to as "craftsmanship snobbery," suggests that all you need for success in agile development is a small team of really good people. However, in this case, a team with a spread of talent was able to produce world-class results.

Frequent Releases Build Trust

Reducing WIP shortens lead time. Shorter lead times mean that it is possible to release working code more often. More frequent releases build trust with external teams, particularly the marketing team or business sponsors. Trust is a hard thing to define. Sociologists call it social capital. What they've learned is that trust is event driven and that small, frequent gestures or events enhance trust more than larger gestures made only occasionally.

When I teach this in classes, I like to ask women in the class what they think after they go on a first date with a guy. I suggest that they had a nice time and then he doesn't call them for two weeks. He then turns up on their doorstep with a bunch of flowers and an apology. I ask them to compare this to a guy who takes the time to type a text message on his way home that evening to say, "I had a great time tonight. I really want to see you again. Call you tomorrow?" and then follows up by actually calling the next day. Guess who they prefer? Small gestures often cost nothing but build more trust than large, expensive (or expansive) gestures bestowed occasionally.

And so it is with software development. Delivering small, frequent, high-quality releases builds more trust with partner teams than putting out larger releases less often.

Small releases show that the software development team can deliver and is committed to providing value. They build trust with the marketing team or business sponsors. High quality in the released code builds trust with downstream partners such as operations, technical support, field engineering, and sales.

Tacit Knowledge

It's quite easy to speculate why small batches of code improve quality. Complexity in knowledge-work problems grows exponentially with the quantity of work-in-progress. Meanwhile, our human brains struggle to cope with all this complexity. So much of knowledge transfer and information discovery in software development is tacit in nature and is created during collaborative working sessions, face-to-face. The information is verbal and visual but it's in a casual format like a sketch on a whiteboard. Our minds have a limited capacity to store all this tacit knowledge and it degrades while we store it. We fail to recall precise details and make mistakes. If a team is co-located and always available to each other, this memory loss can be corrected through repeated discussion or tapping the shared memory of a group of people. So agile teams co-located in a shared workspace are more likely to retain tacit knowledge longer. Regardless of this, tacit knowledge depreciates with time, so shorter lead times are essential for tacit knowledge processes. We know that reducing work-in-progress is directly related to reducing lead times. Hence, we can infer that there will be lower tacit-knowledge depreciation when we have less work-in-progress, resulting in higher quality.

In summary, reducing work-in-progress improves quality and enables more frequent releases. More frequent releases of higher quality code improve trust with external teams.

Balance Demand against Throughput

Balancing demand against throughput implies that we will set the rate at which we accept new requirements into our software development pipe to correspond with the rate at which we can deliver working code. When we do this, we are effectively fixing our work-in-progress to a given size. As work is delivered, we will pull new work (or requirements) from the people creating demand. So any discussion about prioritization and commitment to new work can happen only in the context of delivering some existing work.

The effect of this change is profound. The throughput of your process will be constrained by a bottleneck. It's unlikely you know where that bottleneck is. In fact, if you speak to everyone in the value stream, they will probably all claim to be completely overloaded. However, once you balance demand against throughput and limit the work-in-progress within your value stream, magic will happen. Only the bottleneck resources will remain fully loaded. Very quickly, other workers in the value stream will find they have slack capacity. Meanwhile, those working in the bottleneck will be busy, but not swamped. For the first time, perhaps in years, the team will no longer be overloaded and many people will experience something very rare in their careers, the feeling of having time on their hands.

Create Slack

Much of the stress will be lifted off the organization and people will be able to focus on doing their jobs with precision and quality. They'll be able to take pride in their work and will enjoy the experience all the more. Those with time on their hands will start to put that time toward improving their own circumstances; they may tidy up their workspace or take some training. They will likely start to apply themselves to bettering their skills, their tools, and how they interact with others up- and downstream. As time passes and one small improvement leads to another, the team will be seen as continuously improving. The culture will have changed. The slack capacity created by the act of limiting work-in-progress and pulling new work only as capacity is available will enable improvement no one thought was possible.

You need slack to enable continuous improvement. You need to balance demand against throughput and limit the quantity of work-in-progress to enable slack.

Intuitively, people believe they have to eliminate slack. So after limiting work-in-progress by balancing demand against throughput, the tendency is to "balance the line" by adjusting resources so that everyone is efficiently fully utilized. Although this may look efficient and satisfy typical twentieth-century management accounting practices, it will impede the creation of an improvement culture. You need slack to enable continuous improvement. In order to have slack, you must have an unbalanced value stream with a bottleneck resource. Optimizing for utilization is not desirable.

Prioritize

If the first three steps in the recipe have been implemented, things will be running smoothly. High-quality code should be arriving frequently. Development lead times should be relatively short, as work-in-progress is limited. New work should be pulled in to development only as capacity is freed up on completion of existing work. At this point, management's attention can turn to optimizing the value delivered rather than merely the quantity of code delivered. There is little point in paying attention to prioritization when there is no predictability in delivery. Why waste effort trying to order the input when there is no dependability in the order of delivery? Until this is fixed, management time is better used to focus on improving both the ability to deliver and the predictability of delivery. You should turn your thoughts to ordering the priority of the input once you know you can actually deliver things in approximately the order they are requested.

Influence

Prioritization should not be controlled by the engineering organization and hence is not under the control of engineering management. Improving prioritization requires the product owner, business sponsor, or marketing department to change their behavior. At best, engineering management can seek only to influence how prioritization is done.

In order to have political and social capital to influence change, a level of trust must have been established. Without the capability to deliver high-quality code regularly, there can be no trust and hence little possibility to influence prioritization and thus optimize the value being delivered from the software team.

Recently, it's become popular in the Agile community to talk about business-value optimization and how the production rate of working code (called the "velocity" of software development) is not an important metric. This is because business value delivered is the true measure of success. While this may ultimately be true, it is important not to lose sight of the capability maturity ladder that a team must climb. Most organizations are incapable of measuring and reporting business value delivered. They must first build capability in basic skills before they try greater challenges.

Building Maturity

This is how I think a team should mature: First, learn to build high-quality code. Then reduce the work-in-progress, shorten lead times, and release often. Next, balance demand against throughput, limit work-in-progress, and create slack to free up bandwidth, which will enable improvements. Then, with a smoothly functioning and optimizing software development capability, improve prioritization to optimize value delivery. Hoping for business value optimization is wishful thinking. Take actions to get to this level of maturity incrementally—follow the Recipe for Success.

Attack Sources of Variability to Improve Predictability

Both the effects of variability and how to reduce it within a process are advanced topics. Reducing variability in software development requires knowledge workers to change the way they work—to learn new techniques and to change their personal behavior. All of this is hard. It is therefore not for beginners or for immature organizations.

Variability results in more work-in-progress and longer lead times. This is explained more fully in chapter 19. Variability creates a greater need for slack in non-bottleneck resources in order to cope with the ebb and flow of work as its effects manifest on the flow of work through the value stream. A full understanding of why this is true requires some background in statistical process control and queuing theory, which is beyond the scope of this book. Personally, I like the work of Donald Wheeler and Donald Reinertsen on variability and queuing; so, if you want more information on those topics, start there.

For now, take it on trust that variability in the size of requirements, and in the amount of effort expended on analysis, design, coding, testing, build integration, and delivery adversely affect the throughput of a process and the costs of running a software development value stream.

However, some sources of variability are inadvertently designed into processes through poor policy choices. The case study in chapter 4 highlights several examples: the monthly replanning, the service-level agreement on estimations, and the priority of the production text changes. All three of these examples are controlled by policies that can be changed. Simply changing an existing process policy can dramatically reduce sources of variability that affect predictability.

Recipe for Success and Kanban

Kanban enables all six steps in the Recipe for Success. Kanban delivers on the Recipe for Success and the Recipe for Success delivers on its promise for the manager implementing it. In turn, the Recipe for Success illustrates why Kanban is such a valuable technique.

Takeaways

❖ Kanban delivers all aspects of the Recipe for Success.

❖ The Recipe for Success explains why Kanban has value.

❖ Poor quality can represent the largest waste in software development.

❖ Reducing work-in-progress improves quality.

❖ Improved quality improves trust with downstream partners such as operations.

❖ Releasing frequently improves trust with upstream partners such as marketing.

❖ Demand can be balanced against throughput with a pull system.

❖ Pull systems expose the bottlenecks and create slack in non-bottlenecks.

❖ Good quality prioritization maximizes the value delivered by a well-functioning software development value chain.

❖ Prioritization is of little value without good initial quality and predictability of delivery.

❖ Making changes to reduce variability requires slack.

❖ Reducing variability reduces the need for slack.

❖ Reducing variability enables a resource balancing (and, potentially, a reduction in headcount).

❖ Reducing variability reduces resource requirements.

❖ Reducing variability allows reduced kanban tokens, less WIP, and results in reduced average lead time.

❖ Slack enables improvement opportunities.

❖ Process improvement leads to greater productivity and greater predictability.

❖ Chapter 4 ❖

From Worst to Best in Five Quarters

In October 2004, Dragos Dumitriu was a program manager at Microsoft. He had recently taken charge of a department that had a reputation as the worst in Microsoft's IT division.

The title "Program Manager" at Microsoft is more readily interpreted elsewhere as project manager, but there it typically also includes some responsibility for analysis and architecture. A program manager is assigned to some initiative, project, or product and has responsibility for a feature or set of features. A program manager will recruit resources from functional areas such as development and testing in order to complete the work. In Dragos's case, he was responsible for the software maintenance for the XIT business unit. This team (shown in Figure 4.1), based at a CMMI Model Level 5 rated vendor

Figure 4.1 The team in Hyderabad, India, in late 2004; Dragos is fourth from the left.

in India, and consisting of 3 developers, 3 testers and local managers, developed minor upgrades and fixed production bugs for about 80 cross-functional IT applications used by Microsoft staff throughout the world. At this time, I was also working there.

The Problem

Dragos had volunteered to take charge of the team that had the worst reputation for customer service within Microsoft. His job as change agent, determined to fix the long lead times and poor expectation-setting, was hindered by the political climate. Several of his predecessors in this position were still colleagues working on other projects within the same business unit, and they worried that were he to improve the performance, it would make them look bad in comparison.

The programmers and testers working for the vendor were following the Software Engineering Institute's Personal Software Process/Team Software Process (PSP/TSP) methodology. Microsoft mandated this contractually. Jon De Vaan, who at that time reported directly to Bill Gates, was a big fan of Watts Humphrey of the Software Engineering Institute. As head of Engineering Excellence at Microsoft he was in a position to mandate processes used within the IT department and by their vendors. This meant that changing the software development lifecycle method in use was not an available option.

Dragos realized that neither the PSP/TSP method nor the CMMI rating of the vendor were likely to be the root cause of their problems. In fact, the team produced pretty much what was asked of them, and with very high quality. Nevertheless, they had a five-month lead time on change requests and this, along with their backlog of requests, was growing uncontrollably. The perception was of a team that was badly organized and managed. As a result, senior management was not disposed to provide additional money to fix the problem.

So the constraints on change were political, fiscal, and company-policy related. He asked for my advice.

Visualize the Workflow

I asked Dragos to sketch the workflow. He drew a simple stick-man drawing describing the lifecycle of a change request, and as he did so, we discussed the problems. Figure 4.2 is a facsimile of what he drew. The PM stick figure represents Dragos.

Requests were arriving uncontrollably. Four product managers represented and controlled budgets for a number of customers who owned applications maintained by XIT. They were adding new requests, including escaped production defects (discovered in

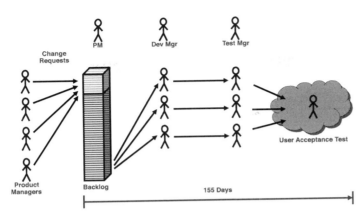

Figure 4.2 XIT Sustaining Engineering: Initial workflow showing lead time

the field). These defects had not been created by the maintenance team, but by the application development project teams. Those application development teams were generally broken up one month after the release of a new system, and the source code was handed off to the maintenance team.

Factors Affecting Performance

When each request arrived, Dragos would send it to India for an estimate. The policy was that estimates had to be made and returned to the business owners within 48 hours. This would facilitate making some return-on-investment (ROI) calculation and deciding whether to proceed with the request. Once a month, Dragos would meet with the product managers and other stakeholders, and they would reprioritize the backlog and create a project plan from the requests.

At this time, the number of monthly throughput requests was around seven. The backlog had 80 or more items in it and it was growing. This meant that 70 or more requests were being reprioritized and rescheduled each month, and that requests were taking more than four months, on average, to process. This was the root cause of the dissatisfaction. These requests were small, and the constant reprioritization meant that requestors were being continually disappointed.

The requests were tracked with a tool called Product Studio. An updated version of this tool was later released publicly as "Team Foundation Server Work Item Tracking." The XIT maintenance team was a type of organization I see often in my teaching and consulting work—they had lots of data, but they were not using it. Dragos began to mine the data and discovered that an average request took 11 days of engineering.

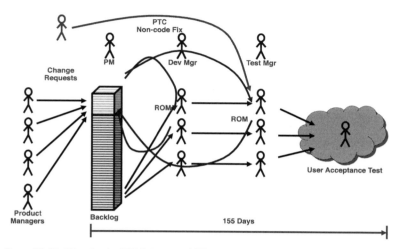

Figure 4.3 Workflow showing ROM Estimates and PTC input

However, lead times of 125 to 155 days were typical. More than 90 percent of the lead time was queuing, or other forms of waste.

The estimates for new incoming work were consuming a lot of effort. We decided to analyze this using some guesswork. Despite being referred to as "rough order of magnitude" (ROM) estimates, the customer expectation was actually for a very accurate estimate, and team members had learned to take great care over preparing them. Each one was taking about one day for each developer and tester. We quickly calculated that the estimation effort alone was consuming around 33 percent of capacity, and on a bad month it could be as much as 40 percent. This capacity was allocated in preference to working on coding and testing. Estimating new requests was also apt to randomize plans made for that month.

In addition to change requests, the team had a second type of work, known as production text changes (PTCs), that were generally graphical or textual in nature, or involved modifying values in tables or XML files. These changes did not require a developer and were often made by business owners, product managers, or the program manager, but they did require a formal test pass, so they affected the testers.

PTCs arrived without much warning and were, by tradition, expedited over all other work or estimation effort. PTCs tended to arrive in sporadic batches. They also would randomize any plans made for that month (see Figure 4.3).

Make Process Policies Explicit

The team was following the required process that included many bad policy decisions that had been made by managers at various levels. It is important to think of a process as a set of policies that govern behavior. These policies are under management's control. For example, the policy to use PSP/TSP was set at the executive vice-president level, one rung below Bill Gates, and this policy would be hard or impossible to change. However, many other policies, such as the policy to prioritize estimates over actual coding and testing, were developed locally and were under the collaborative authority of the immediate managers. It is possible that the policies made sense at the time they were implemented; but circumstances had changed, and no attempt had been made to review and update the policies that governed the team's operation.

Estimation Was a Waste

After some discussion with his colleagues and manager, Dragos decided to enact two initial management changes. First, the team would stop estimating. He wanted to recover the capacity wasted by estimation activity and use it to develop and test software. Eliminating the schedule randomization caused by estimating would also improve predictability, and the combination would, he hoped, have a great impact on customer satisfaction.

However, removing estimation was problematic. It would affect the ROI calculations, and customers might worry that bad prioritization choices were being made. In addition, estimates were used to facilitate inter-departmental cost accounting and budget transfers. Estimates were also used to implement a governance policy. Only small requests were allowed through system maintenance. Larger requests, those exceeding 15 days of development or testing, had to be submitted to a major project initiative and go through the formal program management office (PMO) portfolio management governance process. We will revisit these issues shortly.

Limit Work-in-Progress

Note: This is a policy choice. One change request per developer at any given time is a policy. It can be modified later. Thinking of a process as a set of policies is a key element of the Kanban Method.

The other change Dragos decided to make was to limit work-in-progress and pull work from an input queue as current work was completed. He chose to limit WIP in development to one request per developer and to use a similar rule for testers. He inserted a queue between development and test in order to receive the PTCs and to smooth the flow of work between development and test, as shown in Figure 4.4. This approach of using a buffer to smooth out variability in size and effort is discussed in chapter 19.

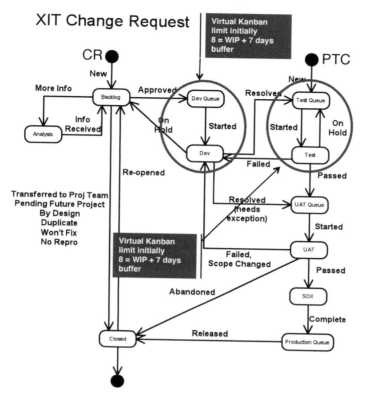

Figure 4.4 State chart modeling desired workflow with WIP limits

Establishing an Input Cadence

Note: Cadence is a concept in the Kanban Method that determines the rhythm of a type of event. Prioritization, delivery, retrospectives, and any recurring event can have its own cadence.

In order to facilitate the decision to limit WIP and institute a pull system, Dragos had to think about the cadence for interacting with the product managers. He thought that a weekly meeting would be feasible, as the topic of the meeting would be the simple replenishment of the input queue from the backlog. In a typical week there might be three slots free in the queue. So the discussion would center around the question, "Which three items from the backlog would you most like delivered next?" This cadence is modeled in Figure 4.5.

He wanted to offer a "guaranteed" delivery time of 25 days from acceptance into the input queue. This 25 days was considerably greater than the 11 days of average engineering time required to complete the job. The statistical outliers required around 30 days, but he anticipated very few of them; 25 days sounded attractive, especially

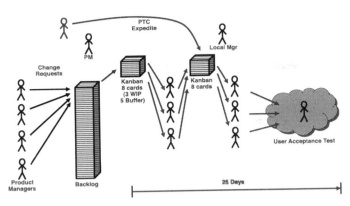

Figure 4.5 Workflow with Kanban WIP limits and queues

compared to the existing lead time of around 140 days. He expected to hit that target with regularity, building trust with product managers and their customers as he went.

Striking a New Bargain

Dragos took an offer to the product managers. He asked them to accept that they would meet once per week to discuss prioritization, that he would limit work-in-progress, and that his team would stop estimating. In exchange for this, he would guarantee delivery within 25 days and that he'd report due-date performance against that metric.

So the customers were being asked to give up ROI calculations and inter-department budget transfers based on estimated effort per request. In exchange, they were being offered unprecedented short delivery times and reliability. To get around the accounting issues, they were asked to accept that all requests took, on average, 11 days of engineering, as observed from historical data and described on page 37. They were asked to accept that costs were essentially fixed. They were being asked to ignore the cost-accounting paradigm on which their inter-department budget transfers were based.

The explanation to justify this was that the vendor had a 12-month contract that was invoiced monthly. The vendor allocated people against the contract and those people were paid regardless of whether or not they were working. The budget to fund this from the four product managers was simply a fixed amount, allocated proportionately. Dragos guaranteed that each of the product managers would get a fair allocation of capacity. This would free them from tracking individual requests. If they could accept that their dollars bought them capacity and that capacity was guaranteed, maybe they could ignore their bias for unit-based costing and budget transfers. Some simple rules were created to determine who should pick to refill the queue so that capacity was allocated fairly. A simple, weighted round-robin scheme was sufficient to achieve this.

Implementing Changes

While the product managers and many of Dragos's management colleagues in the XIT unit remained skeptical, the consensus was that he should give these changes a try. After all, things were bad and getting worse. This could surely not make it any worse than it already was! Someone had to try something, and Dragos had been expected to implement some changes.

So the changes were enacted.

It began to work. Requests were processed and released to production. Delivery times on new commitments were met within the 25-day promise. The weekly meeting worked smoothly, and the queue was replenished each week. Trust began to build with the product managers.

Adjusting Policies

You may wonder how prioritization was facilitated if there was no longer an ROI calculation performed. It turned out that it wasn't needed. If something was important and valuable, it was selected for the input queue from the backlog; if it wasn't, then it wasn't selected. Some time later Dragos recognized that a new policy was needed: Any item older than six months was purged from backlog. If it wasn't important enough to be selected within six months of its arrival, it could be assumed that it wasn't important at all. If it truly was important, it could be resubmitted.

What was there about the governance policy to prevent large items from sneaking through maintenance when they should be submitted instead to a major project? This was solved by accepting that some might sneak through. The historical data told us that these were less than two percent of requests. Developers were instructed to be alert, and if a new request they started on appeared to be large, and they estimated that it required greater than 15 days of effort, then they should alert their local manager. The risk and cost of doing this was less than one-half of one percent of available capacity.

> **Note:** This is a common theme in the Kanban Method. The combination of explicit policies, transparency, and visualization empowers individual team members to make their own decisions and to manage risks themselves. Management comes to trust the system because they understand that the process is made of policies. The policies are designed to manage risk and deliver customer expectations. The policies are explicit, work is tracked transparently, and all team members understand the policies and how to use them.

It was a great tradeoff. By dropping estimates, the team gained more than 33 percent of capacity at the risk of less than 1 percent of that capacity. This new policy empowered developers to manage risk and to speak up when necessary!

The first two changes were left to settle in for six months. A few minor changes were made during this period. As mentioned, a backlog purge policy was added; the weekly meeting with product owners also disappeared. The process was running so smoothly that Dragos had the Product Studio tool modified so it would send him an email when a slot became available in the input queue. He would then alert the product owners via email, who would decide among themselves who should pick next. A choice would be made and a request from the backlog was replenished into the queue within two hours of a slot becoming available.

Looking for Further Improvements

Dragos began looking for further improvement. He'd been studying historical data for tester productivity from his team and comparing it with other teams within XIT supplied by the same vendor. He suspected that the testers were not heavily loaded and had a lot of slack capacity. By implication the developers were a significant bottleneck. He decided to visit the team in India. On his return he instructed the vendor to make a headcount-allocation change. He reduced the test team from three to two and added another developer (Figure 4.6). This resulted in a near-linear increase in productivity, with the throughput for that quarter rising from 45 to 56.

Microsoft's fiscal year was ending. Senior managers were noticing the significant improvement in productivity and the consistency of delivery from the XIT Sustained

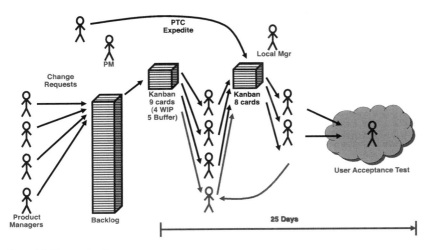

Figure 4.6 Resource leveling

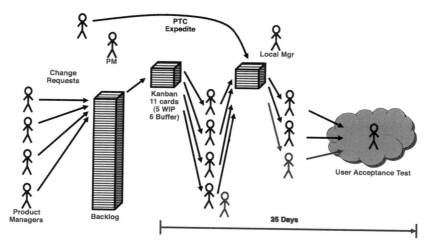

Figure 4.7 Adding additional resources

Engineering (software maintenance) team. Finally, management trusted in Dragos and the techniques he was employing. The department was allocated enough money for two more people: One developer and one additional tester were added in July 2005. The results were significant (Figure 4.7).

Results

The additional capacity was enough to increase throughput beyond demand. The result? The backlog was eliminated entirely on November 22, 2005. By this time the team had reduced the lead time to an average of 14 days against an 11-day engineering time. The due-date performance on the 25-day delivery time target was 98 percent. The throughput of requests had risen more than threefold, while lead times had dropped by more than 90 percent, and reliability improved almost as much. No changes were made to the software development or testing process. The people working in Hyderabad were unaware of any significant change. The PSP/TSP method was unchanged and all the corporate governance, process, and vendor-contract requirements were fully met. The team won the Engineering Excellence Award for the second half of 2005. Dragos was rewarded with additional responsibilities, and the day-to-day management of the team was handed off to the local line manager in India, who relocated to Washington.

These improvements came about in part because of the incredible managerial skill of Dragos Dumitriu, but the basic elements of mapping a value stream, analyzing flow,

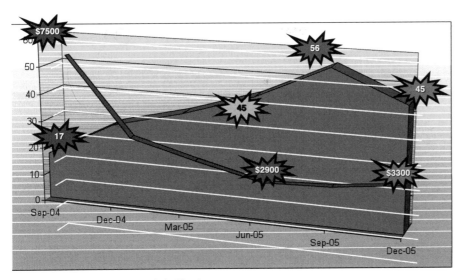

Figure 4.8 Quarterly throughput overlaid with unit cost

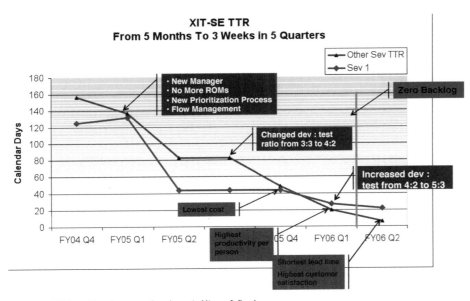

Figure 4.9 XIT Sustaining time to resolve, shown in Microsoft fiscal years

setting WIP limits, and implementing a pull system were key enablers. Without the flow paradigm and the kanban approach of limiting WIP, the performance gains would not have been possible. Kanban enabled incremental changes with low political risk and low resistance to change.

The XIT case shows how a WIP-limited pull system was implemented on a distributed project using offshore resources, and facilitated with an electronic tracking tool. There was no visual board and many of the more sophisticated features of the Kanban Method described in this book had yet to emerge. Nevertheless, what manager could ignore the possibility that a greater than 200 percent productivity improvement, with a 90 percent lead-time reduction, could be achieved with significantly improved predictability and with minimal political risk and resistance to change?

Takeaways

❖ The first kanban system was implemented with the XIT Sustained Engineering software maintenance team at Microsoft, starting in 2004.

❖ The first kanban system used an electronic tracking tool.

❖ The first kanban system was implemented with an offshore team at a CMMI Model Level 5 appraised vendor in Hyderabad, India.

❖ The workflow should be sketched and visualized.

❖ The process should be described as an explicit set of policies.

❖ Kanban enables incremental changes.

❖ Kanban enables change with reduced political risk.

❖ Kanban enables change with minimal resistance.

❖ Kanban will reveal opportunities for improvement that do not involve complex changes to engineering methods.

❖ The first kanban system produced a greater than 200-percent productivity boost, a 90-percent lead time reduction, and a similar improvement in predictability.

❖ Significant improvements are possible by managing bottlenecks, eliminating waste, and reducing the variability that affects customer expectations and satisfaction.

❖ Changes can take time to take full effect. This first case study took 15 months to enact.

❖ **Chapter 5** ❖

A Continuous Improvement Culture

In Japanese, the word *kaizen* literally means "continuous improve-ment." A workplace culture where the entire workforce is focused on continually improving quality, productivity, and customer satisfaction is known as a "kaizen culture." Very few businesses have actually achieved such a culture. Companies like Toyota, where employee participation in their improvement program is close to 100 percent, and where, on aver-age, each employee gets one suggestion implemented every year as part of on-going improvement, are very rare.

In the software development world, the Software Engineering Institute (SEI) of Carnegie Mellon University has defined the high-est level of their Capability Maturity Model Integration (CMMI) as Optimizing. Optimizing implies that the quality and performance of the organization is continuously being refined. While not explicitly stated, as the CMMI says little about culture, achieving optimizing behavior as an organization is more likely to happen within a *kaizen* culture.

Kaizen Culture

To understand why it is so hard to achieve a kaizen culture, we must first understand what such a culture would look like. Only then can we discuss why we might want to achieve such a culture and what its benefits might be.

In kaizen culture the workforce is empowered. Individuals feel free to take action; free to do the right thing. They spontaneously swarm on problems, discuss options, and implement fixes and improvements. In a kaizen culture, the workforce is without fear. The underlying norm is for management to be tolerant of failure if the experimentation and innovation was in the name of process- and performance improvement. In a kaizen culture, individuals are free (within some limits) to self-organize around the work they do and how they do it. Visual controls and signals are evident, and work tasks are generally volunteered for rather than assigned by a superior. A kaizen culture involves a high level of collaboration and a collegial atmosphere where everyone looks out for the performance of the team and the business above themselves. A kaizen culture focuses on systems-level thinking while making local improvements that enhance overall performance.

A kaizen culture has a high level of social capital. It is a highly trusting culture where individuals, regardless of their position in the decision-making hierarchy of the business, respect each other and each person's contribution. High-trust cultures tend to have flatter structures than lower-trust cultures. It is the degree of empowerment that enables a flatter structure to work effectively. Hence, achieving a kaizen culture may enable elimination of wasteful layers of management and reduce coordination costs as a result.

Many aspects of a kaizen culture are in opposition to established cultural and social norms in modern Western culture. In the West, we are brought up to be competitive. Our school systems encourage competition in academics and in athletics. Even our team sports tend to encourage the development of heroes and teams built around one or two exceptionally talented players. The social norm is to focus on the individual first and to rely on outstanding individuals to deliver victory or to save us from peril. It is little wonder that we struggle in the workplace to encourage collegial behavior and systems-level thinking and cooperation.

Kanban Accelerates Organizational Maturity and Capability

The Kanban Method is designed to minimize the initial impact of changes and reduce resistance to adopting change. Adopting Kanban should change the culture of your organization and help it mature. If the adoption is done correctly, the organization will morph into one that adopts change readily and becomes good at implementing changes and process improvements. The SEI refers to this as a capability at Organizational Innovation and Deployment (OID)[15] within the CMMI model. It has been shown[16] that organizations that achieve this high level of capability in change management can adopt Agile development methods, such as Scrum, faster and better than less mature organizations.

When you first implement Kanban you are seeking to optimize existing processes and change the organizational culture rather than switch out existing processes for others that may provide dramatic economic improvements. This has led to the criticism[17] that Kanban merely optimizes something that needs to be changed. However, there is now considerable empirical evidence[18] that Kanban accelerates the achievement of high levels of organizational maturity and capability in core high-maturity process areas such as Causal Analysis and Resolution (CAR) and Organizational Innovation and Deployment.

When you choose to use Kanban as a method to drive change in your organization, you are subscribing to the view that it is better to optimize what already exists, because that is easier and faster and will meet with less resistance than running a managed, engineered, named-change initiative. Introducing a radical change is harder than incrementally improving an existing one. You should also understand that the collaborative-game aspects of Kanban will contribute to a significant shift in your corporate culture and its maturity. This shift will later enable much more significant changes, again with lower resistance, than if you were to try to make those changes immediately. Adopting Kanban is an investment in the long term capability, maturity, and culture of your organization. It is not intended as a quick fix.

Case Study: Corbis Application Development

When I introduced a kanban system at Corbis, in 2006, I did so for many of the mechanical benefits that were demonstrated with Microsoft XIT in 2004 (as described in chapter 4). The initial application was the same—IT applications maintenance. I was not anticipating a significant cultural shift or a significant shift in organizational maturity. I did not expect what we now know as the Kanban Method to evolve from this work.

As I write this book in 2010, it now has been established that Kanban is a natural fit for IT maintenance work. Back in 2006 it wasn't yet clear, but a kanban system form seemed to fit well with the functional problems of maintenance work. I didn't go to Corbis with the intent of "doing Kanban." I did go there with the intent of improving customer satisfaction with the software development team. It was a happy coincidence that the first problem to be addressed was the lack of predictability concerning delivery from IT software maintenance.

Background and Culture

In 2006, Corbis was a privately held business, with around 1,300 employees worldwide. It controlled the digital rights to many fascinating works of art as well as representing approximately 3,000 professional photographers, licensing their work for use by publishers and advertisers. It was the second-largest stock photography company in the world. There were other lines of business, too, the most notable of which was the rights-licensing business that controlled the rights, on behalf of families, estates, and management firms, to the images and names of personalities and celebrities. The IT department consisted of about 110 people split between software engineering and network operations/systems maintenance. The workforce was augmented from time to time with contract staff to work on major projects. At its peak in 2007, the software engineering department employed 105 people, including 35 contingent staff in Seattle and another 30 at a vendor in Chennai, India. Most of the testing was performed by this team in Chennai. There was a very traditional approach to project management: Everything was planned in a dependency tree of tasks and rolled up by a program-management office. It was a company with a conservative culture, in what had been a relatively conservative and slow-moving industry. It employed conservative, traditional approaches to project management and the software engineering lifecycle.

The IT department maintained a diverse set of approximately 30 systems. Some were fairly typical accounting and human-resource systems; others were exotic, and at times esoteric, applications for the digital-rights management industry. There was a wide range of technologies, software platforms, and languages supported. The workforce was incredibly loyal; many people in the IT department had been with the company more than eight years, some with as many as 15 years of service. Not bad for a company that was about 17 years old. The existing process was a traditional, waterfall-style software development lifecycle (SDLC) that had been institutionalized over the years with the creation of a business analysis department, a systems analysis department, a development department, and an offshore testing department. Within these departments there were many specialists, such as analysts whose background was accounting and whose specialty was finance. Some developers were also specialists, for example, J.D. Edwards' programmers, who maintained the J.D. Edwards accounting software.

None of this was ideal; but it was what it was. Things were the way they were. When I joined the company there was some expectation and trepidation that I would impose an Agile method and use my positional power to force people to change their behavior. Although this might have worked, it would have been brutal, and the impact during the transition would have been severe. I was afraid of making things worse; afraid that projects would grind to a halt while new training was provided and staff adapted to new ways of working. I was also afraid of losing key personnel, knowing that the workforce was fragile due to the excessive levels of specialization. I chose to introduce a kanban system, get the systems-maintenance work back on track, and to see what happened from there.

The need for a Software Maintenance Function

Software maintenance (or "RRT" for "Rapid Response Team," as it was known internally) had been funded by the executive committee with an additional ten percent budget for the software engineering department. This equated to five additional people in 2006. Some time before my arrival, those five people had been hired. Due to the diverse nature of the systems involved and the existing high degree of specialization on the team, it was decided that a dedicated team of five people to do maintenance work would not be a good solution. So five additional people were added to the general pool of resources: one project manager, one analyst, one developer, and two testers. This introduced an additional complication that it was necessary, from a governance perspective, to show that the additional five people were actually doing maintenance work and hadn't simply been sucked into the major-project portfolio. However, on any given day, those five people could be any of the approximately 55 people on the team.

One solution would have been for everyone to complete complex timesheets, which would have added an administrative burden, to show that 10 percent of the team's hours were being spent on maintenance activities. This would have been highly intrusive, but it is typical of how middle managers respond to such a challenge. Another approach was to introduce a kanban system.

An expectation had been set that a maintenance team would enable Corbis to make incremental releases to IT systems every two weeks. Major projects had typically involved major system updates and new systems releases once every three months. But as the business matured and the nature of these systems became more complex, this cadence of quarterly major releases had become intermittent. In addition, some of the existing systems were effectively end-of-life and were really due for complete replacement. Legacy-system replacement is a major challenge, and it typically involves long projects with a large staff until a parity of functionality is reached and the old system can be turned off as the new one is brought online. (This approach is really far from optimal, but it is typical.)

So the maintenance releases were the one area within Corbis IT where kanban could enable some form of business agility.

Small Projects for Maintenance Wasn't Working

The existing system to deliver maintenance releases, the system that was broken, was to schedule a series of short, two-week-long projects. This might have been recognizable as Agile software development using two-week iterations, but it wasn't. When I first arrived, the negotiation of scope for a two-week release cycle was taking about three weeks. So the front-end transaction costs of a release were greater than the value-added work. It was taking about six weeks to get a two-week release out the door.

Implementing Change

It was clear before making any changes that the status quo was unacceptable. The current system was unable to deliver the required level of business agility. System maintenance gave us an ideal opportunity to introduce change. Maintenance work was generally not mission critical. It was nevertheless highly visible, as the business had direct input to the prioritization, and the choices were highly tactical and important to short-term business goals. System maintenance was something everyone cared about and wanted to work effectively. And finally, there was a compelling reason to make changes. Everyone was unhappy with the existing system. The developers, testers, and analysts were all aggravated with the time wasted negotiating scope, and the business people were hugely dissatisfied with the results.

We designed a kanban system with scheduled bi-weekly releases, planned for 1 p.m. every second Wednesday, and with scheduled prioritization meetings with the business people, set at 10 a.m. every Monday. So the prioritization cadence was weekly, and the release cadence was bi-weekly. The choice of cadence was determined through collaborative discussions with upstream and downstream partners and based on the transaction and coordination costs of the activities. A few other changes were made. We introduced an Engineering Ready (input) queue with a WIP limit of five items and then added WIP limits throughout the lifecycle of analysis, development, build, and system test. Acceptance test, staging, and ready for production were left unlimited, as they were not considered capacity constrained and were, to some extent, outside our immediate political control.

Primary Effects of the Changes

The effects of introducing a kanban system were, at one level, unsurprising, yet at another, they were quite remarkable. We started to make releases every two weeks. After about three iterations, these were happening without incident. The quality was good and there was little to no need for emergency fixes when the new code went live in production. The overhead for scheduling and planning releases had dropped dramatically, and the bickering between the development teams and the program management office had almost completely disappeared. So kanban had delivered on its basic promise. We were putting out high-quality releases very regularly, with a minimum of management overhead. Transaction and coordination costs of a release had been drastically reduced. The team was getting more work done and they were delivering that work to the customer more often.

It was the secondary effects that were all the more remarkable.

Unanticipated Effects of Introducing Kanban

For the development team, we introduced the physical card wall using sticky notes on a whiteboard in January 2007. We started holding morning standup meetings around the board for 15 minutes at 9:30 a.m. each day. The physical board had a huge psychological effect compared to anything we got from the electronic tracking tool we used at Microsoft. By attending the standup each day, team members were exposed to a sort of time-lapse photography of the flow of work across the board. Blocked work items were marked with pink tickets, and the team became much more focused on issue resolution and maintaining flow. Productivity jumped dramatically.

With the flow of work now visible on the board, I started to pay attention to the workings of the process. As a result, I made some changes to the board. My team of managers came to understand the changes I was making and why I was making them, and by March, they were making changes themselves. In turn, their team members, the individual developers, testers, and analysts, started to see and understand how things worked. By early summer, everyone on the team felt empowered to suggest a change, and we'd observed the spontaneous affiliation of (often cross-functional) groups of individuals who would discuss process problems and challenges and make changes as they saw appropriate. Typically, they would inform the management chain after the fact. What had emerged over approximately six months was a kaizen culture in the software engineering team. Team members felt empowered. Fear had been removed. They took pride in their professionalism and in their achievements and they wanted to do better.

Sociological Change

Since the Corbis experience, there have been other, similar reports from the field. Rob Hathaway of Indigo Blue was the first truly to replicate these results with the IT group at IPC Media in London. The fact that others have been able to replicate the sociological effects of Kanban we observed at Corbis makes me believe that there is a causation and that it was neither a coincidence nor a direct effect of my personal involvement.

I've thought a lot about what brought about these sociological changes. Agile methods have offered us transparency on work-in-progress for a decade, and yet teams following the Kanban Method appear to achieve a kaizen culture faster and more effectively than typical Agile software development teams. Often, teams adding Kanban to their existing Agile methods find a significant improvement in social capital among team members. Why could this be?

My conclusion is that Kanban provides transparency into the work, but also into the process (or workflow). It provides visibility into how the work is passed from one group to another. Kanban enables every stakeholder to see the effects of his or her actions or inactions. If an item is blocked and someone is capable of unblocking it, Kanban shows this. Perhaps there is an ambiguous requirement. Typically, the subject-matter expert who can resolve the ambiguity might expect to receive an email with a request for a meeting. After a follow up call, they arrange a meeting to suit their calendar, perhaps three weeks out. With Kanban and the visibility it provides, the subject-matter expert realizes the effect of inaction and prioritizes the meeting, perhaps rearranging his or her calendar to schedule a meeting this week rather than delay for a further two weeks.

In addition to the visibility into process flow, work-in-progress limits also force challenging interactions to happen sooner and more often. It isn't easy to ignore a blocked item and simply work on something else. This "stop the line" aspect of Kanban seems to encourage swarming behavior across the value stream. When people from different functional areas and with different job titles swarm on a problem and collaborate to find a solution, thus maintaining the flow of work and improving system-level performance, the level of social capital and team trust increases. With higher levels of trust engendered through improved collaboration, fear is eliminated from the organization.

Work-in-progress limits coupled with classes of services (explained in chapter 11) also empower individuals to make scheduling decisions on their own, without management supervision or direction. Empowerment improves the level of social capital by demonstrating that superiors trust subordinates to make high-quality decisions on their own. Managers are freed up from supervising individual contributors and can focus their mental energy on other things, such as process performance, risk management, staff development, and improved customer and employee satisfaction.

Kanban greatly enhances the level of social capital within the team. The improved levels of trust and the elimination of fear encourage collaborative innovation and problem solving. The net effect is the rapid emergence of a kaizen culture.

Viral Spread of Collaboration

Kanban clearly improved the atmosphere in the software engineering department at Corbis, but it was the results beyond that group that were the most remarkable. How the viral spread of Kanban improved collaboration around the company is worth reporting and analyzing.

Case Study: Corbis Application Development, continued

Each Monday morning at 10 a.m., Diana Kolomiyets, the project manager responsible for coordinating the IT systems maintenance releases would convene the RRT prioritization board meeting. The business attendees were typically vice presidents. They ran a business unit and reported to a senior vice president or C-level officer of the company. Put another way, a vice president reported to an executive-committee member. Corbis was still small enough that it made sense for such a high-ranking manager to attend the weekly meeting. Equally, the tactical choices being made were sufficiently important that they really needed the direction of a vice president to influence a good choice.

Usually, each attendee received an email on the Friday prior to the meeting. It would state something like, "We anticipate that there will be two slots free in the queue next week. Please examine your backlog items and select candidates for discussion at Monday's meeting."

Bargaining

In the first few weeks of the new process, some of the attendees would come with an expectation of negotiating. They might say, "I know there is only one slot free, but I have two small ones, can you just do them both?" This bargaining was rarely tolerated. The other members of the prioritization board ensured that everyone played by the rules. They might reply, "How do I know they are small? Should I take you at your word?" or counter with, "I've got two small ones too. Why shouldn't I get my favorites selected?" I refer to this as the "Bargaining Period," as it indicates the style of negotiation that took place at prioritization meetings.

Democracy

After about six weeks, and coincidentally around the same time that the development team introduced the use of the physical whiteboard, the prioritization board introduced a democratic voting system. They spontaneously volunteered this, as they'd become tired of bickering. The bargaining at the meeting was wasting time. It took a few iterations to refine the voting system, but it settled down to a system where each attendee got one vote for each free slot in the queue that week. At the beginning of the meeting, each member would propose a small number of candidates for selection. As time went by, proposing requests got more sophisticated; some people came with PowerPoint slides, others with spreadsheets that laid out a business case. Later we heard that some members were lobbying their colleagues by taking them to lunch. Deals were being done, "If I vote for your choice this week, will you vote for my choice next week?" Underlying the new democratic system of prioritization, the level of collaboration between business units at the vice-president level was growing. Although we didn't realize it at the time, the level of social capital across the whole firm was growing. When leaders of business units start collaborating, so, it seems, do the people within their organizations. They follow the lead from their leader. Collaborative behavior coupled with visibility and transparency breeds more collaborative behavior. I refer to this period as the "Democracy Period."

Down with Democracy

Democracy was all very well, but after a further four months, it seemed that democracy had failed to elect the best candidate. A considerable effort was expended implementing an e-commerce feature for the Eastern European market. The business case had been stellar but its candidacy was suspect from the beginning, and some had questioned the quality of the data in the business case. After several attempts, this feature had been selected and was duly implemented. It was one of the larger features processed through the RRT system, and many people got involved and noticed it. Two months after launch, our Director of Business Intelligence did some data mining on the revenue generated. It was a fraction of what had been promised in the original business case, and the estimated payback period against the effort expended was calculated at 19 years. Due to the transparency that Kanban offered us, many stakeholders became aware of this, and there was discussion about how precious capacity had been wasted on this choice when a better choice might have been made instead. That was the end of the Democracy Period.

Collaboration

What replaced it was quite remarkable. Bear in mind that the prioritization board consisted mostly of vice president–level employees and officers of the company. They had broad visibility into aspects of the business that many of us were unaware of. So at the beginning of the meeting, they started to ask, "Diana, what is the current lead time for delivery?" She might reply, "Currently we are averaging 44 days into production." So then they asked a simple question: "What is the most important tactical business initiative in this company 44 days out from now?" There might be some discussion, but typically there was swift agreement. "Oh, that'll be our European marketing campaign launching at the conference in Cannes." "Great! What items in the backlog are required to support the Cannes event?" A quick search might produce a list of six items. "So, there are three slots free this week. Let's pick three from six and we'll get to the others next week." There was very little debate. There was no bargaining or negotiation. The meeting was over in about 20 minutes. I've come to refer to this as the "Collaboration Period." It represents the highest level of social capital and trust between business units that was achieved during my time as Senior Director for Software Engineering at Corbis.

Cultural Change is Perhaps the Biggest Benefit of Kanban

It was interesting to see this cultural change emerge and to see how it affected the wider company as employees followed the lead of their vice presidents and started to collaborate more with their colleagues from other business units. This change was sufficiently profound that the recently appointed chief executive, Gary Shenk, called me to his office to ask if I had any explanation. He told me that he had observed a new level of collaboration and collegial spirit in the senior ranks of the company and that formerly antagonistic business units seemed to be getting along a lot better. He suggested that the RRT process had something to do with it and asked if I had any explanation for it. While I am sure that I wasn't as articulate as I am now, writing this chapter two years later, I convinced him that our kanban system had greatly enhanced collaboration and with that, the level of social capital among everyone involved.

The cultural side effects of what we now recognize to be capital-K Kanban were quite unexpected and in many ways counter-intuitive. He asked, "Why aren't we doing all our major projects this way?" Why indeed? So we set about implementing Kanban in the major-project portfolio. We did this because Kanban had enabled a kaizen culture, and that cultural change was so desirable that the cost of changing the many mechanics of prioritization, scheduling, reporting, and delivery that would result from implementing Kanban was considered a price worth paying.

Takeaways

- ❖ *Kaizen* means "continuous improvement."

- ❖ A kaizen culture is one in which individuals feel empowered, act without fear, affiliate spontaneously, collaborate, and innovate.

- ❖ A kaizen culture has a high degree of social capital and trust between individuals, regardless of their level in the corporate hierarchy.

- ❖ Kanban provides transparency on both the work and the process through which the work flows.

- ❖ Transparency of process allows all stakeholders to see the effects of their actions or inactions.

- ❖ Individuals are more likely to give of their time and collaborate when they can see the effect it will have.

- ❖ Kanban WIP limits enable "stop the line" behavior.

- ❖ Kanban WIP limits encourage swarming to resolve problems.

- ❖ Increased collaboration from swarming on problems and interaction with external stakeholders raises the level of social capital within the team and the trust among team members.

- ❖ Kanban WIP limits and classes of service empower individuals to pull work and make prioritization and scheduling decisions without supervision or direction from a superior.

- ❖ Increased levels of empowerment increase social capital and trust among workers and managers.

- ❖ Collaborative behavior can spread virally.

- ❖ Individuals will take their lead from senior managers. Collegial, collaborative behavior among senior leaders will affect the behavior of the whole workforce.

❖ PART THREE ❖

Implementing Kanban

❖ **Chapter 6** ❖

Mapping the Value Stream

Kanban is an approach that drives change by optimizing your existing process. The essence of starting with Kanban is to change as little as possible. You must resist the temptation to change workflow, job titles, roles and responsibilities, and specific working practices. Everything from which the team members and other partners, participants, and stakeholders derive their self-esteem, professional pride, and ego should be left unchanged. The main target of change will be the quantity of WIP and the interface to and interaction with upstream and downstream parts of your business. So you must work with your team to map the value stream as it exists. Try not to change it or invent it in an idealistic fashion.

In some political situations, there may be an official process that is not being followed. When you attempt to map the value stream, your team will insist that you re-document the official process, not the actual process being used. You must resist this and insist that the team map the process they actually use. Without this, it will be impossible to use a card wall as a process-visualization tool because team members can use the card wall only if it reflects what they actually do.

Defining a Start and End Point for Control

It is necessary to decide where to start and end process visualization, and in doing so, define the interface points with upstream and downstream partners. It's important to handle this early-stage Kanban implementation sensitively, as making poor choices early on may invoke failure. Successful teams have tended to stick to adopting workflow visualization with cards and limiting WIP within their own political sphere of control, and negotiating a new way of interacting with immediate upstream and downstream partners. For example, if you control the engineering or software development function and have control or influence over analysis, design, testing, and coding, then map this value stream and negotiate new styles of interaction with the business partners upstream who provide requirements, prioritization, and portfolio management, and those downstream with system operations, or production-maintenance functions. By drawing the boundaries this way, you are asking your own team to adopt only the behavior of limiting WIP. You are not asking up- or downstream teams to change the way they do their jobs. You are not asking them to limit WIP and implement a pull system. However, you are asking them to interact differently with you—to interact in a way that is compatible with the pull system you want to implement.

Work Item Types

Once you have selected the starting point in the workflow or value stream, identify the types of work that arrive at that point and any others that exist within the workflow that will need to be limited. For example, bugs are likely a type of work that exists within the workflow. You may also identify other types of development-centric work such as refactoring, systems maintenance, and infrastructure upgrades and related rework. For incoming work, you may have types like User Story or Use Case or Functional Requirement or Feature. In some cases, the incoming types might be hierarchical, such as Epic, a collection of user stories.

Typical work item types seen on teams adopting Kanban have included, but are not limited to, the following.

- Requirement
- Feature
- User Story
- Use Case
- Change Request

- Production Defect

- Maintenance

- Refactoring

- Bug

- Improvement Suggestion

- Blocking Issue

It is useful to name work item types for their source, such as Regulatory Requirement, Field Sales Request, Strategic Planning Requirement, and so forth. Using a naming convention that makes the source of the work request transparent provides additional context and enables the system to evolve to serve multiple customers.

Work item types will tend to be defined by the source of the work, the flow of the work, or the size of the work. For example, the PTCs from the Microsoft example in chapter 4 have a different workflow, though the source is the same as the Change Requests. It doesn't make sense to have separate kanban systems for the two types. The same team does the work. It is easy enough to visualize the types by using different colors of ticket or different rows (swim lanes) on a card wall. Order of magnitude in size might typically be: small (a few days), medium (a week or two), or large (a month or more). Each order of magnitude should have its own type.

Drawing a Card Wall

It's typical to draw card walls to show the activities that happen to the work rather than specific functions or job descriptions. Often there is a strong overlap between a function and an activity; for example, analysts perform analysis. However, the convention with Kanban in software projects over the last few years has been to model the work and not the workers, the functions, or the handoffs between functions.

Before drawing a card wall to visualize workflow, it may make sense to sketch or model it. Figure 4.4 (page 40) shows a very formal model, using state chart notation, of the desired workflow, with queues added for change requests and production text changes processed by the sustaining engineering team of XIT at Microsoft. You may find a less formal approach is perfectly adequate. A stick-man drawing, similar to those shown throughout chapter 4, or a flow chart or its equivalent may suffice.

Once you have understood your workflow by sketching or modeling it, start to define a card wall by drawing columns on the board that represent the activities performed, in

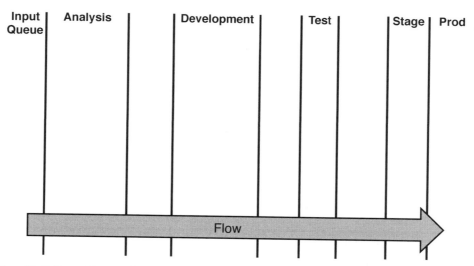

Figure 6.1 Outline workflow on card wall (flow left to right)

the order they are performed, as shown in Figure 6.1. When drawing columns initially, it makes sense to draw these with a marker. However, through usage the lines will be erased. During the first few weeks you may find that you'll want to make a few changes to the workflow, so continuing to use an erasable marker makes sense. However, there will be a point when you will want something more permanent. Very narrow rolls of vinyl tape are available from office-supply stores, specifically designed for precision work on whiteboards, as illustrated in Figure 6.2. At Corbis it became commonplace to delineate columns and rows on a card wall using this tape. This practice is now widely adopted, with teams using various grades and widths of tape to mark rows and columns.

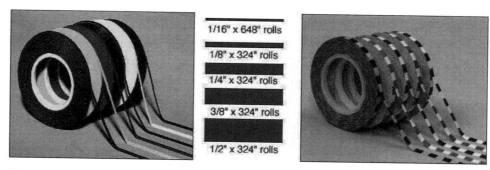

Figure 6.2 Precision whiteboard tape

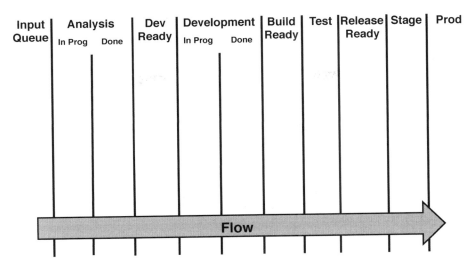

Figure 6.3 Workflow with added buffers and queues

Note that for activity steps it is necessary to model both the in-progress and the completed work; by convention this is done by splitting the column.

Next, add the input queue and any downstream delivery steps that you wish to visualize, as shown in Figure 6.3.

Finally, add any buffers or queues that you believe are necessary. There are some differing schools of thought on this, and it is really an advanced topic. A full discussion of where to put buffers and how to size them is beyond the scope of this book, so it will suffice for now to describe two popular approaches:

- The first school of thought says do not try to second-guess the location of bottleneck or the source of variability that will require a buffer. Rather, implement the system and wait for the bottleneck to reveal itself, then make changes to introduce a buffer.

  A variant on this suggests that WIP limits should be set fairly loosely initially so that variability, waste, and bottlenecks do not have a significant impact on the pull system when it is first implemented. This is discussed more fully in chapter 10 and later, in chapters 17 and 19.

- Another school of thought takes a different approach. It suggests that rather than implement loose WIP limits to avoid a challenging introduction of the system, each stage should be buffered, and the activity steps should have tight limits. Bottlenecks and variability will reveal themselves by how full the buffers become. Small, simple changes can then be made to reduce buffer sizes and then eventually you can eliminate unnecessary buffers.

Figure 6.4 Card wall illustrating use of diamond-shaped tickets at the top of queuing or buffer columns (Courtesy of Liquidnet Holdings, Inc.)

At the time of this writing, there isn't enough evidence to suggest which approach is better.

Some teams have adopted a convention of showing buffers and queue columns by using a 45-degree rotated card. This provides a strong visual indicator of how much work is flowing rather than queuing at any given instant in time. This allows the team and other stakeholders to "see," literally, the amount of economic cost (or waste) in the system.

Demand Analysis

For each type of work identified, you should make a study of the demand. If you have historical data, use it to make a quantitative study. If you do not, then an anecdotally derived subjective analysis will suffice. For example, in the Microsoft XIT example from chapter 4, there were two types of work, Change Requests and Production Text Changes (PTCs). Arguably, the Change Requests should have been broken out further into two types, Production Defects and Change Requests (for new functionality). If I were coaching this team today, I would recommend that they track four types of work in total: Change Requests, Production Defects, Production Text Changes, and Bugs (or un-escaped defects).

For each of these types we would study the demand. The demand for PTCs came in bursts. There might be no PTCs for six weeks, then in a single week there would be a burst of perhaps ten arriving almost at once. PTCs were small and fast to implement. This meant that their impact was not severe. Designing a system to cope with intermittent demand like this is difficult. If PTCs had represented a significant effort, the system would have required considerable slack built into it in order to adequately cope with PTCs without severely impacting predictability on Change Requests.

Change Requests, on the other hand, arrived at a much steadier pace. While their arrival was stochastic in nature, demand was relatively steady, at perhaps five to seven new requests per week. It would be possible to plot the arrival rate of PTCs in a chart and graph the demand to understand the mean rate of arrival and the spread of variation. The kanban system could then be designed and resourced appropriately to cope with this demand.

Some work item types exhibit seasonal demand, such as regulatory requirements. New tax legislation affects financial and payroll systems in a seasonal fashion. In one case I came across, the IT department of a automobile racing team received regulatory changes from the sport's governing body at the start of each racing season. They may also receive some regulatory requirements during the season, but the volume over the close-season was significantly greater, as the main racing regulations were changed from one year to the next. It is important to understand this demand so that the kanban system design can be adjusted to cope with the demand for different types of work.

Allocating Capacity According to Demand

Once you have an understanding of the demand, you can decide how to allocate capacity within the kanban system to cope with that demand. The example in Figure 6.5 shows three swim lanes, one for each of type of work, namely, change requests; internal maintenance work, such as code refactoring; and production text changes. The allocation is 60 percent to change requests, 10 percent for code refactoring work, and 30 percent for production text changes. Given a demand analysis that shows that production text changes arrive in bursts, the allocation shown suggests that significant slack is being reserved to deal with production text changes urgently on arrival without impacting due dates for other work. The allocation of capacity should be aligned to the risk profile. If, for example, it is acceptable for due-date performance to fall when production text changes arrive, and for lead times on change requests to be both longer and less predictable, the allocation could be different. Perhaps 85 percent for change requests, 10 percent for maintenance, and 5 percent for production text changes. Yet another choice would be to leave a swim lane for production text changes, but not allocate any capacity for it, and adopt a policy of exceeding the work-in-progress limit when a burst of production text changes arrives. This policy will eliminate slack and produce an optimal economic outcome during normal operation. However, when a

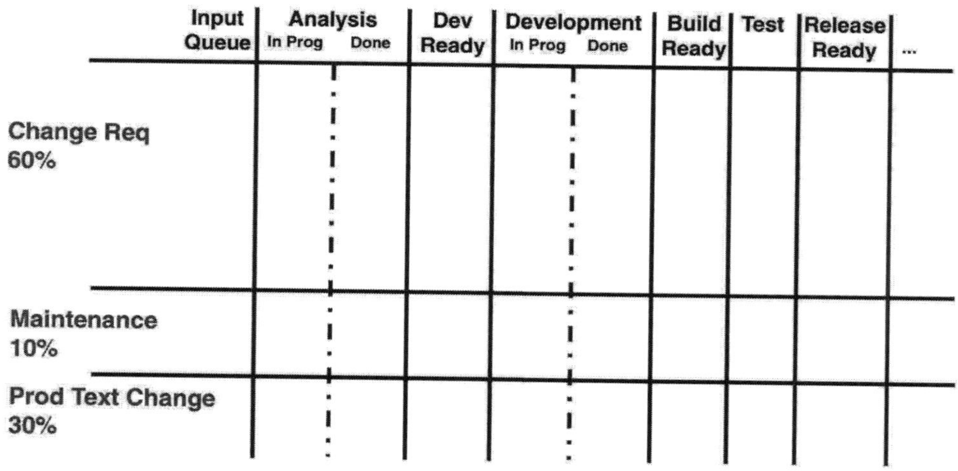

	Input Queue	Analysis		Dev Ready	Development		Build Ready	Test	Release Ready	
		In Prog	Done		In Prog	Done				...
Change Req 60%										
Maintenance 10%										
Prod Text Change 30%										

Figure 6.5 Kanban board with type swim lanes, indicating capacity allocation

burst of production text changes arrives, other work may be severely impacted in both lead time and predictability. This was the choice made in the real example from chapter 4; there was no spare capacity reserved for coping with production text changes.

Later, when we discuss setting work-in-progress limits, we will use this allocation information to set specific limits for the queues in each swim lane.

Anatomy of a Work Item Card

Each visual card representing a discrete piece of customer-valued work has several pieces of information on it. The design of the card is important. The information on the cards must facilitate the pull system and empower individuals to make their own pull decisions. The information on a ticket may vary by work item type or by class of service (discussed in chapter 11).

In the example in Figure 6.6, the number in the top left is the electronic tracking number used to uniquely identify the item and to link it to the electronic version of the tracking system. The title of the item is written in the middle. The date the ticket entered the system is written in the bottom left, which serves a double purpose: It facilitates first-in, first-out (FIFO) queuing for the standard class of service, and it allows team members to see how many days have expired against the service-level agreement (described in chapter 11). For fixed delivery date class-of-service items, the required delivery date is written in the bottom right of the ticket.

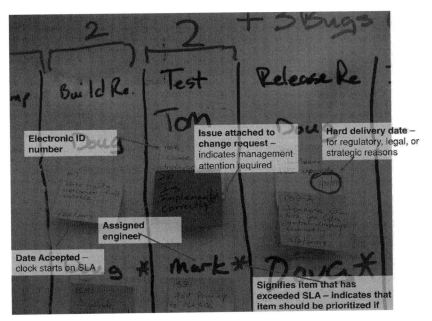

Figure 6.6 Close up of card wall showing anatomy of work item cards

In the example shown in Figure 6.6, some other information is shown off-ticket. A star designates that the item is late against the target lead time in the service-level agreement. More recently, I've seen this achieved by attaching a sticker to the upper-right-hand corner of the ticket. The name of the assigned person is also written above the ticket. Because the assigned person will change as the ticket flows across the board, it is undesirable to write a name on the ticket. However, more recent implementations feature small name tags stuck to the items, the use of magnets (where the whiteboard is magnetic), and stickers or magnets that feature avatars of team members. Characters from *South Park* are a popular choice of avatar. Any mechanism that allows team members and immediate management to see at a glance who is working on what will suffice.

As a general rule, the design of the ticket used to represent an individual piece of work should have sufficient information to facilitate project-management decisions, such as which item to pull next, without the intervention or direction of a manager. The idea is to empower team members with transparency of process, project goals and objectives, and risk information. As you discover more about classes of service and service-level agreements, you'll discover that Kanban facilitates a powerful self-organizing risk-management mechanism. Equally, Kanban, by empowering team members to make their own scheduling and prioritization decisions, shows respect for individuals and a trust in the system (or process design.) A well-designed work item card is a key enabler of a high-trust culture and a Lean organization.

Electronic Tracking

Electronic tracking has been a feature of kanban systems in software development since they were first introduced in 2004. It is, however, optional. For teams that are distributed geographically, or those who have policies that allow team members to work from their homes one or more days per week, electronic tracking is essential. Electronic tracking can be performed with basic ticketing and work item tracking systems such as Jira, Microsoft Team Foundation Server, Fog Bugz, and HP Quality Center. However, a more powerful system will allow you to visualize the work item tracking as if it were on a card wall.

At the time of this writing, a number of web-based and application-based tools are emerging in the marketplace to provide electronic tracking using visual boards that simulate card walls with columns, WIP limits, and other essential aspects of Kanban. These include, but are not limited to: Lean Kit Kanban, Agile Zen, Target Process, Silver Catalyst, RadTrack, Kanbanery, VersionOne, Greenhopper for Jira, Flow.io, and some additional open-source projects that add Kanban interfaces to tools such as Team Foundation Server and FogBugz. Figure 6.7 shows an example from AgileZen.

Electronic tracking is necessary for teams that aspire to higher levels of organizational maturity. If you anticipate the need for quantitative management, organizational process performance (comparing the performance across kanban systems, teams or projects), and/or causal analysis and resolution (root-cause analysis based on statistically sound data), you will want to use an electronic tool from the beginning.

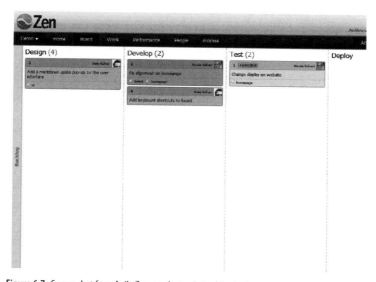

Figure 6.7 Screen shot from Agile Zen, an electronic tracking tool

Setting Input and Output Boundaries

Align the design of the kanban system and card wall with the decision made earlier to limit the boundary of WIP control. It is likely that upstream and downstream partners will ask later to visualize their work on your card wall. However, it is better to provide transparency onto your own work first and wait for others to ask to be part of your Kanban initiative.

In the example shown in Figure 6.8, the input queue is designated "E.R." for "Engineering Ready." It made sense to set the input point at this step in the lifecycle because the upstream business analysis department reported up through a different part of the organization structure. There was little trust or collaboration between the managers across the two groups. The input queue was, therefore, replenished from the backlog of requirements generated by the business analysis department.

In this example, the downstream handoff is the deployment to production. Once the software is deployed and handed off the to the network and systems operations department for daily maintenance and support, it is considered out-of-scope.

Coping with Concurrency

One common occurrence when designing a card wall for a kanban system is a process in which two or more activities can happen concurrently, for example, software

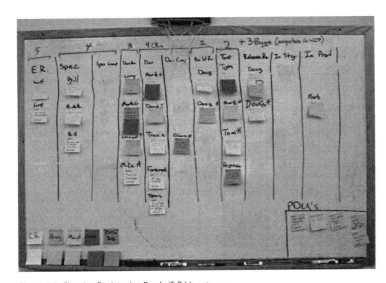

Figure 6.8 Showing Engineering Ready (E.R.) input queue

development and test development. There are two basic patterns for coping with this situation. One is not to model it at all; just leave a single column where both activities can occur together (Figure 6.9). This is simple, but many teams have not preferred it. Some teams have adapted this model by using different colors or shapes of ticket to show the different activities.

The other option is to split the board vertically into two (or more) sections (Figure 6.10).

In this example, some tagging mechanism to tie items in the top and bottom of the board is necessary. This may simply involve use of, for example, the top right-hand corner of the ticket to cross-reference the associated item. In a good electronic tracking system it is possible to link related items such as development and test activities.

Coping with Unordered Activities

Particularly in highly innovative and experimental work, there may be several activities that need to happen with a piece of customer-valued work, but those activities do not need to happen in any particular order. In these circumstances, it is important to realize that Kanban should not force you to complete the activities in a given order.

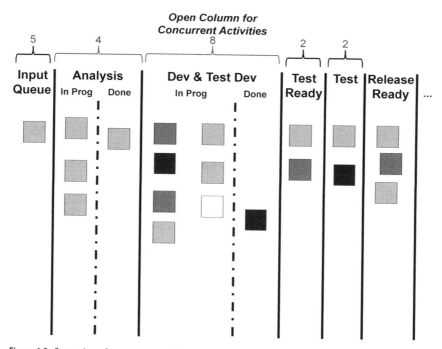

Figure 6.9 Open column for concurrent activities

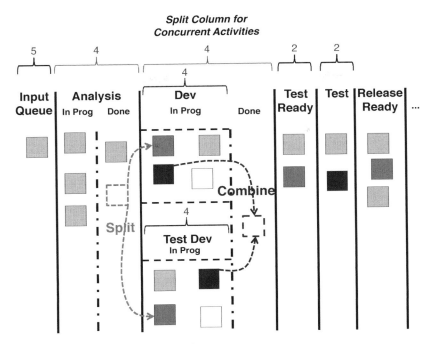

Figure 6.10 Split column for concurrent activities

What is most important when modeling your kanban system is that it must reflect the way the real work is done.

There are a couple of strategies to the multiple unordered-activities problem. The first is similar to coping with concurrency: Simply have a single column as a bucket for the activities and do not explicitly track on the board which of them is complete.

The second, and potentially more powerful choice, is to model the activities in a similar fashion to the concurrent activities. In this design, as shown in Figure 6.11, the tickets have to move vertically up and down the column as they are pulled into each of the specific activities. Visualizing which activities have been completed on each item can be done by modifying the ticket design to have a small box for each activity. When the activity is complete, the box can be filled to visually signal that the item is ready to be pulled to another activity in the same column. If all the boxes are filled, the item is ready to be pulled to the next column on the board, or it can be moved to a "done" column (Figure 6.12).

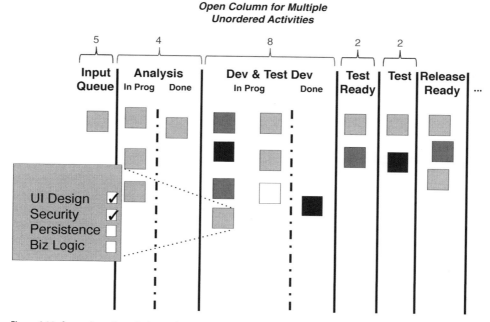

Figure 6.11 Open column for mulitple unordered activities

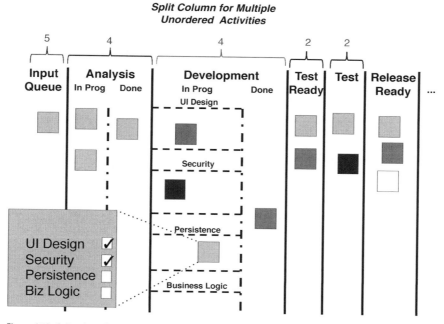

Figure 6.12 Split column for mulitple unordered activities

Takeaways

❖ Decide the outer boundaries of the kanban system. It is often best to limit this to the immediate span of political control. Do not force visualization, transparency, and WIP limits on any department that does not volunteer to collaborate.

❖ Model the card wall to align with the boundary decisions made regarding limiting WIP and visualizing work.

❖ Define work item types and model how their work flows. Some types may not require every step in the flow.

❖ Design the work item card to have enough information to facilitate self-organization for pull and to enable team members to make good quality decisions with respect to risk based on work item type, service-level agreements, and classes of service.

❖ Use an electronic tracking system if the team is distributed, has some work-from-home policy, or aspires to higher maturity behaviors that require the quantitative information that an electronic system can provide

❖ Where appropriate, discuss methods for handling concurrency in activities and choose how to model and visualize them.

❖ Where appropriate, discuss methods for handling activities that do not need to follow a specifically ordered flow and choose how to model and visualize them.

❖ **Chapter 7** ❖

Coordination with Kanban Systems

Visual Control and Pull

When people talk about Kanban, the most popular form of coordination that comes to mind is a card wall. Typically, the work-in-progress limits are drawn on the board at the top of each column or across a span of columns. Pull is signaled if the number of cards in a column is less than the indicated limit. In Figure 7.1, we can see that the limit written above Analysis is four items. However, there are only three cards in that column. Because 4 – 3 = 1, this signals that we can pull one item into Analysis (the systems-analysis function) from the input queue, Engineering Ready (marked "E.R." in Figure 7.1). In turn, the input queue has a limit of five. There are currently only two items remaining in that queue. When we pull one into Analysis, there will be only one item remaining (5 – 1 = 4). This signals that we can prioritize four new items into the input queue at the next prioritization meeting.

When the team decides to pull an item, they can choose which item to pull based on available visual information, such as the work item type, the class of service, the due date (if applicable), and the age of the work item. Policies for pull related to class of service are discussed in chapter 11.

Figure 7.2 shows a close-up of sticky notes that represent work items on our card wall. Color is being used to communicate a combination

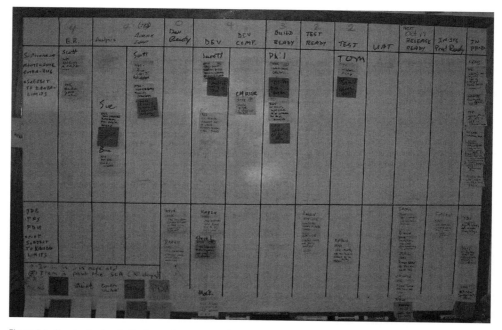

Figure 7.1 Showing kanban limits above columns on a card wall

of work item type and class of service. The name of the owner or assigned staff member is written above the card. Some teams like to use additional, smaller sticky notes with names, or small avatars stuck to the work item to signify who is working on it. This allows everyone on the team to see who is working on what.

In figure 6.6, the electronic tracking number is shown in the top left-hand corner of the sticky note. The date an item entered the input queue is shown in the bottom left. The age of the item can be deduced from this date. If an item is of a class of service that is a guaranteed delivery date, that is shown in the bottom right. If an item is late, that is signified with a red star above the card at top right. If something is blocked, a pink issue ticket is attached to the blocked item. In the example in Figure 7.2, the issue is a first-class work item in its own right, hence has its own electronic tracking number, a date when it entered the system, and the assigned staff member's name shown above it.

This scheme is idiosyncratic to the first kanban implementation at Corbis. Your implementation will almost certainly differ. However, you are likely to want to visually capture the assigned staff member, the start date, the electronic tracking number, the work item type, the class of service, and some status information, such as whether the item is late. The goal is to visually communicate enough information to make the system self-organizing and self-expediting at the team level. As a visual control

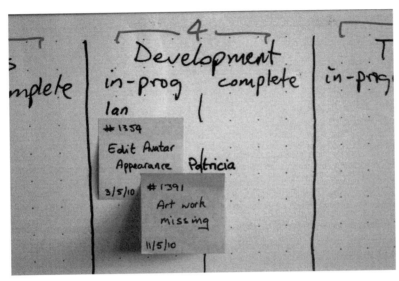

Figure 7.2 Close-up of card wall showing issue ticket attached to blocked item

mechanism, the kanban board should enable team members to pull work without direction from their manager.

Electronic Tracking

As an alternative, or, as a supplement, to a card wall, an electronic system is often used to track work in a kanban system. Some of the tools available for this are listed in chapter 6. For a more up-to-the-minute list, check the Limited WIP Society web site, http://www.limitedwipsociety.org/.

With my team, we implemented our own application, Digital Whiteboard (see Figure 7.3), on top of Team Foundation Server. In the case study in chapter 4, the electronic tracking was done using a Microsoft-internal tool called Product Studio. This was a forerunner to Team Foundation Server, and since 2005 Microsoft has been using Team Foundation Server for its own internal development project tracking.

The application shown in Figure 7.3 shows the kanban limits grouped above columns. It is capable of visually showing when a kanban limit has been exceeded. It also displays the number of status items for each work item, including different icons that show whether an item is late or is blocked with an issue.

Electronic tracking is important for kanban systems because it permits several things that are not viable with a simple card wall. Electronic tracking allows data-gathering

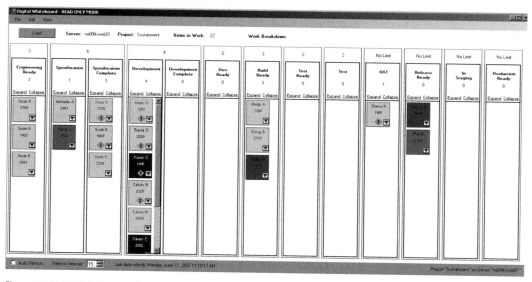

Figure 7.3 Digital Whiteboard application used at Corbis

that can be used to generate metrics and reports both for day-to-day management and for retrospective use, for example, at monthly operations reviews.

Daily Standup Meetings

Standup meetings are a common element of Agile development processes. They typically take place in the morning before work starts and have a generally agreed-upon format. A typical standup meeting is for a single team of up to twelve people—usually about six. The format generally involves working around the assembled group and asking three questions: What did you achieve yesterday? What will you do today? Are you blocked or do you need assistance? Each team member answers these questions and then the team is coordinated to do its work for the day.

Standup meetings have evolved differently with Kanban. The need to go around the room and ask the three questions is obviated by the card wall. The wall contains all the information about who is working on what. Attendees who come regularly can see what has changed since yesterday and whether something is blocked or not is visually evident. So standups take a different format with a kanban system. The focus is on flow of work. The facilitator, typically a project manager or a line manager, will "walk the board." The convention has developed to work backward—from right to left (in the direction of pull)—through the tickets on the board. The facilitator might solicit a status update on a ticket or simply ask if there is any additional information that is not on the board and may not be known to the team.

Particular emphasis will be placed on items that are blocked (have a pink ticket attached) or delayed due to defects (have a series of blue tickets attached). Questions may also be asked about items that appear to be stuck and have not moved for a few days. Some teams have devised ways for visualizing this. For example, at an Italian automobile racing team and sports-car manufacturer they mark a dot beside the ticket for each day that it stays in a single location. This allows the team to question whether an item should be marked as blocked if it is not actively flowing. This improves the issue-management capability of the organization (described more fully in chapter 20.) The team will discuss briefly who is working an issue and when it will be resolved. There will also be a call for any other blocking issues that are not on the board and for anyone who needs help to speak up. More advanced and mature teams will find that they do not need to walk through every card on the wall. They will tend to focus only on tickets that are blocked or have defects. This mechanism allows standup meetings to scale to much larger numbers of people; Daniel Vacanti ran a successful standup with more than 50 people at a project at Corbis in 2007 where, despite the large size of team, the meeting was completed in around 10 minutes each morning.

The After Meeting

The After Meeting consists of huddles of small groups of 2 or 3 people. This emerged as spontaneous behavior because team members wanted to discuss something on their minds: perhaps a blocking issue, perhaps a technical design or architecture issue, but more often, a process-related issue. The After Meeting is a vital element of the cultural transformation that emerges with Kanban. After Meetings generate improvement ideas and result in process tailoring and innovation.

On larger projects, some After Meetings took the form of established Scrum-style standup meetings. Teams of up to six people working together on a feature, story, or requirement would meet briefly to coordinate their efforts for the day. There is an interesting difference between this emergent Kanban process behavior and Scrum. With Scrum, the teams meet first and then send a delegate to a Scrum-of-Scrums to coordinate a program or large project. In Kanban the behavior is reversed—the program-level meeting happens first.

Queue Replenishment Meetings

Queue replenishment meetings serve the purpose of prioritization in Kanban. This prioritization is said to be deferred or postponed until the last reasonable moment due to the nature of the queue-replenishment mechanism and the cadence of the meetings. Queue replenishment meetings are held with a group of business representatives or product owners (to use popular Agile development vernacular). It's recommended that these meetings happen at regular intervals. Providing a steady cadence for queue replenishment reduces the coordination cost of holding the meeting and provides certainty and reliability over the relationship between the business and the software development team.

The purpose of such a meeting is to fill the kanban system's input queue for a single value stream, system, or project. Stakeholders who have an interest in the deliveries from the team and who have items waiting in the backlog should attend this meeting. The business attendees should be as senior in their organizations as possible. More senior people can make more decisions and often have access to wider contextual information. This improves the quality of decision making and optimizes the selection process for queue replenishment.

Ideally, a prioritization meeting will involve several product owners or business people from potentially competing groups within the company. The tension created by this actually becomes a positive influence on good decision making and stimulates a healthy, collaborative environment with the software development team. If only one product owner attends, there is the potential for the interaction to be adversarial.

Other interested stakeholders should be present at the meeting, ideally including: anyone responsible for delivery, e.g., a project manager; at least one technical functional manager, such as a development or test manager, or a more senior technical function manager; some people who can assess technical risk, for example, a technical- or data architect; a usability professional; an operations and systems specialist; and a business analyst. With my team in 2007, one development manager, the manager of the analysis team, and occasionally the enterprise architect or the data architect attended the meeting. The development managers took turns, attending the meeting on a rotating schedule.

The cadence of prioritization meetings will affect the queue sizing in the kanban system and hence the overall lead time through the system. To maximize the agility of the team, it is recommended that the meetings be as frequent as is reasonably possible; weekly is a commonly recommended interval.

Some teams have evolved into demand-driven prioritization rather than use a regular meeting. This is recommended only for more mature organizations in which all stakeholders in the meeting can be available on demand. In the Microsoft case study from chapter 4, the project manager created database triggers that alerted him to a free slot in the input queue. He would then instigate a prioritization discussion with the four product owners via email, alerting them that a slot was available for prioritization. An electronic discussion would ensue and a new item from the backlog would be selected. This process typically took about two hours. By having this on-demand system rather than a weekly meeting, the input queue size could be reduced, which led to a subsequent improvement in lead time through the system.

Release Planning Meetings

Release planning meetings happen specifically to plan downstream delivery. If releases are occurring regularly, with a cadence of, say, bi-weekly, then it makes sense to schedule the release planning activity to take place regularly. This reduces the coordination cost of holding the meeting and ensures that everyone who needs to attend will have the time available.

The person responsible for coordinating the delivery, usually a project manager, typically leads release planning meetings. Any other interested parties should be invited: This usually includes configuration management specialists; systems operation and network specialists; developers; testers; business analysts; and for all of these individual contributors, their immediate supervisors or managers. Specialists are present for their technical knowledge and risk-assessment capabilities. Managers are present so that decisions can be made.

A mature organization will have a checklist or framework for a release that facilitates planning. Some of the things to be considered are the following.

- Which items in the system are (or will be) ready for release?
- What is required to actually release each item to production?
- What testing will be required post-release to validate the integrity of production systems?
- What risks are involved?
- How are these risks being mitigated?
- What contingency plans are required?
- Who needs to be involved in the release and present during the push to production (or other delivery mechanism)?

- How long will the release take?

- What other logistics will be involved?

The outcome should be a completed template representing a release plan. With teams that are particularly sophisticated, I have seen the release scripted as an orchestration of procedures to be executed in a given order.

In a large meeting, completing the release plan may not be possible, and some independent follow-up work may be required on the project manager's part.

Triage

Triage is a term borrowed from the medical profession; it refers to the practice of assessing and classifying emergency patients into categories for priority of attention. The system was first used in battlefield medical units, where patients were separated into three categories: beyond help and likely to die soon, likely to live only if given immediate treatment, and likely to survive without immediate treatment. Emergency rooms now use a similar system to prioritize patients as they arrive for treatment.

Triage was adopted into software development for classifying defects (bugs) during the stabilization phase of a traditional software project. Triage is used to classify bugs that will be fixed, and their priority, versus bugs that will not be fixed and will be allowed to escape into production when the product is released. A typical defect triage involves a test lead, a test supervisor or manager, a development lead, a development supervisor or manager, and a product owner.

With Kanban it still makes sense to triage defects. However, the most useful application of triage is to the backlog of items waiting to enter the system.

Backlog triage should be held at relatively infrequent intervals. (Note: some Agile software development methods refer to this as "backlog grooming.") Monthly, quarterly, and twice-yearly are intervals that are popular with teams. The attendees at a backlog triage will typically be the same product owners or business representatives who attend the queue replenishment meeting, along with the project manager. The technical people typically do not attend in such large numbers. Perhaps one technical-function middle manager will be present.

The purpose of a backlog triage is to go through each item on the backlog and decide whether it should remain in the backlog or be deleted. It is not to stack rank or provide any prioritization beyond the simple keep-or-delete choice.

Some teams have avoided the need for triage through automation and policy. The Microsoft XIT team from the chapter 4 case study would delete any item older than six months at a regular monthly interval. The reasoning was that if an item had not been selected for the input queue in six months, it was unlikely to be of significant value and therefore unlikely *ever* to be selected. If this changed, it was also likely that it would be requested again and hence nothing was lost by deleting it from the backlog.

The purpose of triaging the backlog is to reduce its size. The benefit of a smaller backlog is that it facilitates easier prioritization discussions. If there are 200 items in a backlog, it will take significantly less time to pick winners at a prioritization meeting than if there are 2,000 items in the backlog.

A good rule of thumb might be that if the backlog exceeds three months' worth of work, that is, three months of delivery throughput, and all the items in the backlog cannot enter the system within that time, it would be a good idea to prune the backlog. The appropriate size of the backlog will vary according to different markets and domains. Domains with high volatility will want a backlog sized to perhaps one month's worth of items. Domains with very low volatility might be able to hold a backlog with up to one year of items.

Hence, there is a relationship between the size of the backlog, the volatility of the domain in which the individual kanban system is operating, and the delivery velocity, or throughput, of the team. If a team delivers 20 user stories per month and the domain has some, but not excessive, volatility, so that a three-month backlog is desirable, the backlog should have approximately 60 items.

Issue Log Review and Escalation

When work items in the Kanban system are impeded, they will be marked as such and an issue work item will be created. The issue will remain open until the impediment is removed and the original work item can progress through the system. Reviewing open issues, therefore, becomes vital to improving flow through the system.

Issue-log review should happen frequently and regularly. Again, a regular cadence reduces coordination costs and ensures that relevant stakeholders make time to attend. Very mature organizations may be able to dispense with regular meetings and hold on-demand meetings. This would be appropriate if there are a relatively small number of issues occurring and if the increased coordination cost of on-demand meetings is actually less than the cost of holding a regularly scheduled meeting.

Issue-log reviews should involve the project manager plus team members who have logged items as blocked. The main questions to be answered are "Who is assigned to and working the issue?" and "When is the expected resolution?" Issues that are not progressing and are in themselves blocked or stale should be escalated to more senior management.

It may not be necessary to have senior management present at an issue-log review, but it is important to have clearly defined escalation paths and policies. When an issue is blocked, the project manager should take responsibility and escalate the matter appropriately.

Issue management and escalation is typically done very badly, even in Agile development organizations. Resolving issues quickly, particularly issues that are external to the development team, such as environment availability, ambiguous requirements, or lack of test equipment, improves flow and greatly enhances both the team's productivity and the value delivered. Issue management and escalation are core disciplines that provide a big return. Improving them should be a priority even for the most immature teams. This is discussed fully in chapter 20.

Sticky Buddies

The concept of a sticky buddy was introduced at Corbis to resolve a coordination problem. There was a policy that allowed telecommuting at least one day per week, particularly for employees who lived farther out of town than most. The policy dated back to an office move from Bellevue to Seattle, Washington, several years earlier. Personnel telecommuting were able to access the electronic tracking system, version control, build environment, and so forth via VPN. So they were able to see work assigned to them, work on it, complete it, and test it. They were able to update the electronic status of work, marking it complete and available to be pulled downstream. However, they were not physically present in the office and able to move the sticky note on the card wall.

The solution to this was for each person to make a peer-to-peer agreement with someone who would be present in the office to act as their delegate. When the telecommuter completed an item and changed its electronic status, they would contact their sticky buddy by instant message, email, or phone and ask them to update the physical board.

Sticky buddies also facilitated distributed development across different geographic locations. This was particularly important at Corbis, as the test team was in Chennai, India, and there were also some specialist financial-systems developers in Southern California.

Synchronizing across Geographic Locations

Synchronizing teams using kanban systems across multiple geographic locations comes up again and again as an elementary question from those considering implementing a kanban system. Often the questioner assumes that the early Kanban implementations were done in a single geographic location and that I (and other early Kanban advocates) had not considered the challenges of coordinating across geographically distributed teams.

Actually, the opposite is true. The first team at Microsoft, from chapter 4, was actually located in Hyderabad, India, with management and product owners located in Redmond, Washington. Corbis, described in chapter 5, also had people in India and other locations outside Seattle, such as Los Angeles and New York, as well as telecommuters.

The key to coordination across multiple sites is to use an electronic system. It isn't enough to have only a card wall.

In addition to electronic tracking, it will be necessary to keep physical card walls synchronized, at the very least, on a daily basis. It is important to assign someone to take responsibility for this at each location. One team we worked with in 2008 was distributed between New York and Los Angeles. They kept (almost) identical card walls at each location and made a team member responsible for keeping them synchronized on a daily basis.

Some teams also coordinate stand-up meetings over the telephone or by using a video-conferencing system. Prior to any standup meeting, video conference, or telephone call, though, the local responsible person should take time to ensure that the physical board is synchronized with the electronic system.

Takeaways

❖ Best practice is to use both a physical card wall and an electronic tracking system.

❖ Kanban is possible across multiple geographic locations, provided that an electronic tracking system is used.

❖ Electronic systems that simulate the functionality of a physical card wall are available from a variety of vendors.

❖ Holding regular meetings reduces the coordination cost for those meetings and improves attendance.

❖ Prioritization and release planning should be done independently and should have independent cadence.

❖ Daily standup meetings should be used to discuss issues, impediments, and flow. They do not typically follow the established pattern of other Agile development methods.

❖ Daily standups are an essential part of encouraging a culture of continuous improvement. Because the standup brings together the whole team briefly each day, they provide an opportunity for all stakeholders to suggest and discuss improvement opportunities. The period immediately after the standup often develops into an informal process-improvement discussion.

❖ Grooming the backlog with regular triage to reduce its size improves the effectiveness and efficiency of prioritization meetings.

❖ Issue management, escalation, and resolution is a core discipline in improving the performance of a team and should be addressed early in the development of the team.

❖ Escalation paths and policies should be clearly defined.

❖   **Chapter 8**   ❖

Establishing a
Delivery Cadence

Section 3 (chapters 6 through 15) describes the mechanics of implementing a kanban system, concluding with chapter 15, which describes how to get started with a Kanban change initiative. Permission to get started requires striking a different type of bargain with the rest of the external stakeholders than is typical between a software development organization and its partners. Part of this new type of bargain involves agreement and commitment to regular deliveries of working software.

The term "delivery cadence" in the title of this chapter implies establishing a pattern of delivery of working software at a regular interval. For example, if we agreed to make a delivery every two weeks, we'd have a delivery cadence of bi-weekly, or 26 times per year. Perhaps, we would even agree to the day of delivery. For example, every second Wednesday, as was the case for maintenance releases of IT applications at Corbis.

It is generally established in Agile software development circles that a regular cadence is important. Agile development methods achieve this with a time-boxed iteration, typically one week to four weeks in length. The argument for time-boxing is based on the notion that a steady "heartbeat" to a project is important. There is an underlying assumption that in order to achieve this it is necessary to use strictly time-boxed iterations. At the start of the iteration, a scope, or backlog, is agreed upon and a commitment made. Work starts! Some amount of analysis, test planning, design, development, testing, and refactoring is performed. If all goes well, all of the committed scope is completed. The iteration ends with delivery of working software and a retrospective meeting to discuss future improvements and process adjustments. The cycle then begins again. All of this is happening at a regular cadence that has been agreed to in advance—weekly, bi-weekly, monthly, or something else.

Kanban dispenses with the time-boxed iteration and instead decouples the activities of prioritization, development, and delivery. The cadence of each is allowed to adjust to its own natural level. Kanban does not dispense with the notion of a regular cadence, though. Kanban teams still deliver software regularly, preferring a short timescale. Kanban still delivers on the Principles Behind the Agile Manifesto[19]. However, Kanban avoids any dysfunction introduced by artificially forcing things into time-boxes.

Over the last ten years, teams using Agile methods have learned that less WIP is better than more. They've learned that small-batch transfers are better than large ones. As a response to this learning, in the middle part of the last decade, they adopted shorter iteration lengths. Typical Scrum teams went from four weeks to two weeks and Extreme Programming teams from two weeks to one week. One of the problems this introduced is that it can be difficult to analyze work into small enough units to get it done in the available time window. The marketplace responded to this by developing more sophisticated ways for analyzing and writing user stories. The object was to reduce the size of stories to make them more granular and reduce variability in size in order to fit them into smaller iterations. Although this approach is sound in theory, it is hard to achieve. It falls into the category of the sixth element of the Recipe for Success: Attack source of variability to improve predictability. As described in chapter 3, reducing variability often requires people to change their behavior and learn new skills. That means it is hard to do.

So teams have struggled to write user stories consistently small enough to fit into small, time-boxed iterations. This has led to several dysfunctions. The first is to reverse the trend to smaller iterations and go back to larger ones. The alternative is to write stories that are focused on elements of the architecture or some technical decomposition of the requirements. This results in, for example, a story for the user interface, a story for the persistence layer, and so on. A second alternative is to break the story across three iterations in phases, in which the first iteration performs analysis and perhaps test planning, the second involves developing the code, and the third involves system testing and bug fixing. Any or all of these dysfunctions are possible. The latter two make a mockery of the notion of time-boxed iterations and disguise the fact that work is actually still in progress when it is being reported as completed.

Kanban decouples the time it takes to create a user story from the delivery rate. While some work is complete and ready for delivery, some other work will be in progress. Having decoupled lead time for development from delivery cadence, it makes sense to question how often prioritization (and perhaps planning and estimation) should happen. It would seem unlikely that planning, estimation, and prioritization discussions should all need to happen at the same pace as delivery and software release. They are completely different functions, often requiring the attention of different groups of people. The coordination effort around delivery is surely different from the coordination effort around prioritization of new work. Kanban allows the decoupling of these activities.

Kanban also decouples the prioritization cadence from both the lead time through the system and from the delivery cadence. This chapter discusses the elements involved in agreeing on a suitable delivery cadence and when or if it would make sense to have on-demand or ad hoc delivery rather a regularly scheduled delivery. Along the same line of thinking, chapter 9 discusses how to set a prioritization cadence and when or if it would make sense to have on-demand or ad hoc prioritization rather than a regular meeting. And chapter 11 discusses how to set expectations around lead time and how to communicate the contents of a release.

Coordination Costs of Delivery

Coordinating every software delivery has costs. It's necessary to get people together to discuss the deployment (or release), manufacture, packaging, marketing, marketing communications, documentation, end-user training, reseller training, help-desk and technical-support training, installation documentation, installation procedures, staff on-call, on-site schedules during deployment, and so on, and on. Planning the release of a piece of working software can be incredibly complex, depending on the nature of the business domain and the type of software. Upgrading a web site can be quite trivial compared to upgrading firmware deployed in military equipment spread across the globe, or satellites in orbit, or fighter aircraft, or the nodes in a telephone network.

In 2002, when we were planning the release of the PCS Vision upgrade to the Sprint PCS cellular phone network in the United States, tens of thousands of people had to be trained. In stores across the country, 17,000 retail staff had to be trained in the features of the new network and the workings of the 15 or so new handsets being offered. A similar number of people had to be trained to answer the inevitable support calls that would ensue when the unsuspecting public took ownership of their new devices. Just planning the training for around 30,000 people is a major cost in both money and time.

So it is important to understand the coordination costs associated with making a delivery. For example, if software developers have to attend release-coordination meetings, is that distracting them from actually building the software for the release? Following is a list of just some of the questions to consider.

- How many meetings?

- How many people?

- How much time will it consume?

- What opportunity cost is incurred when people are distracted from their regular activities?

Transaction Costs of Delivery

With physical goods, it is easy to understand the transaction costs of making a delivery. First there is payment. The customer will arrange to pay the supplier with some monetary instrument, a credit card, for example. For the pleasure of taking payment via credit card, the leading vendors such as MasterCard and Visa charge the vendor a transaction cost, typically two to four percent of the value of the transaction.

In addition to costs on the financial transaction between the consumer and vendor, there also may be delivery charges. Delivery costs money, but it also costs time and manpower, and there may be installation costs. Say, for example, you buy a washing machine from Sears and arrange for delivery on a given day. Behind the scenes, scheduling the delivery and coordinating with the driver that the correct model is delivered to the correct house at the correct time on the correct day is a coordination cost of delivery. The driver actually picking up the machine at the warehouse, driving it to your home, and unpacking it for you is a transaction cost. Perhaps the same person, or another person, a plumber, installs it for you. The plumber takes time to drive to your home and yet more time to perform the installation. All of this time and effort for delivery and installation is part of the transaction cost of buying that washing machine.

Economically, the retailer absorbs the cost of the credit card transaction. The other transaction costs for delivery and installation are often passed on to the consumer. Not all of the transaction costs are "seen" or "felt" by all the players in the value chain but they affect the economic performance of the system as a whole. The net effect of all these costs is to inflate the final price paid by the consumer without actually increasing the value delivered.

It is true that the washing machine without delivery or installation is of little value, but its value-added capability is that it washes clothes. The delivery and installation are non-value-added activities that should be counted as transaction costs.

In software development, the transaction costs of delivery can also be physical in nature. Some firms, such as Microsoft, still "release to manufacture" (RTM) and print physical media, such as DVDs, and box them up and ship them to distributors, retailers, and other partners. With embedded software, it may be necessary to manufacture a set of chips, or, at least, to "blow" the software code into firmware using technology like EE-PROM. The chips, if necessary, would then need to be physically mounted into the hardware that they control.

In other cases, it may be possible to do an electronic deployment. For example, cell phones now permit what is called over-the-air device management to upgrade the firmware and device settings. Many satellites and space probes allow firmware to be upgraded over the air. This soft deployment capability makes space missions much more agile than they were in the past. The mission can be changed by uploading new software. Defects also can be fixed in situ. Some infamous defects, such as the focusing capability on the Hubble Telescope, were (partially) corrected with software changes. This has changed the economics of deployment.

Many people reading this may be involved in web development or internal application development. Deployment may mean simply copying files across to a set of disks on other machines. While this may sound trivial, it often isn't. Often it is necessary to plan an elaborate procedure to switch off databases, application servers, and other systems gracefully, and then upgrade them and bring them all back again. One of the biggest issues is data migration from one generation of a database schema to another. Databases can get very large. The process of serializing the data to a file, parsing it, unpacking it, perhaps embellishing or augmenting it with other data, then re-parsing it and unpacking it into a new schema can take hours—perhaps even days.

In some environments, software deployment can take hours or days. This is often not because the software is of poor quality or has faulty architecture; it simply reflects the nature of the domain in which the software is used. All the activities involved in successfully delivering software, whether it is packaged applications, embedded firmware, or IT applications running on internal servers, need to be accounted for, planned, scheduled, resourced, and then actually performed. All of these activities are transaction costs of making a delivery.

Efficiency of Delivery

The equation to calculate the efficiency of delivery can be assessed in two ways. The simpler way is to look at the labor and costs involved. The more complex method is to consider the value delivered.

First, the cost-only model. We must consider the total costs incurred between releases. Often this is a known amount—the burn rate of the organization. If we release once per month and our burn rate is $1.3 million per month, our costs are at least $1.3 million per release. We may also incur physical manufacturing costs, printing costs, advertising costs, and out-of-pocket expenditure to coordinate the release. All of this is relatively easily accounted for. Let's imagine it is $200,000 in this case. So our total cost of release is $1.5 million.

We know that our additional out-of-pocket delivery costs are $200,000, but how much of the $1.3 million was also spent planning, coordinating, and actually performing the delivery? If we have suitable time-tracking data available, we might be able to calculate this. Even if we don't, we could make a good guess. How many meetings? How many people? How many hours spent in meetings? Include the number of man-hours for actual deployment or delivery activities. Multiply by the hourly rate. If this added up to $300,000, we'd have a transaction and coordination cost of $500,000 for a delivery.

$$\text{Delivery Efficiency\%} = 100\% \times (\text{Total Cost} - (\text{Coordination Cost} + \text{Transaction Cost})) / \text{Total Cost of Software Release}$$

In this example, our efficiency percentage is

$$100\% \times (\$1,500,000 - \$500,000) / \$1,500,000 = 66.7\%$$

To be more efficient we have to (a) increase the time between deliveries, or (b) reduce the coordination and transaction costs. Choice (a) is the typical choice of twentieth-century Western business. It is the choice that values "economy of scale": Do things in larger batches in order to amortize the costs over longer periods of time. Choice (b) is the typical choice of late twentieth-century Japanese businesses and businesses pursuing Lean Thinking. Choice (b) focuses on reducing of waste by reducing coordination and transaction costs in order to make the batch size efficient—in this case, to make the time between releases efficient.

How much efficiency is enough?

This really is an open question. Each business will have separate views on suitable numbers for efficiency, and a lot will depend on the value to be delivered.

Agreeing a Delivery Cadence

If we understood how much value was to be derived from a given release, we could make a better choice about how frequently to deliver. If our monthly delivery of software were to realize $2 million in revenue versus our costs of $1.5 million, we'd know that we were making $500,000 in margin from the activity. We could rewrite our efficiency equation as

Delivery Efficiency% = 100% × (1 - ((Transaction Costs + Coordination Costs) / (Margin + Transaction Costs + Coordination Costs)))

For our working example, this would produce an efficiency percentage of

$$100\% \times (1\text{-}(\$500{,}000 / (\$500{,}000 + \$500{,}000))) = 50\%$$

Now this gets even more complicated, because calculating the true value of a delivery can be almost impossible. We may not have firm orders at firm prices. We may be speculating on the uptake in the market and the price and margin we can achieve. We may be releasing items of intangible value, such as a revision to our brand identity and marketing materials, or a set of usability and bug fixes to our product or web site.

Calculating whether or not we should delay and release less often to be more efficient is equally difficult. Increasing our time to market may have an adverse effect on our market share, price, and margin. This concept of delivery efficiency is not an exact science. What is most important is that you, the team, and the organization are aware of the costs of making a delivery—in both time and money—and are capable of making some form of rational assessment about the acceptable frequency of delivery.

If it takes ten people three days to make a successful code delivery from a team of fifty people, is it acceptable to make a release every ten working days (or two calendar weeks)? The answer is probably not. Perhaps once per month, or twenty working days, is a better choice. On the other hand, the market might be one where agility and time to market is vital, where a lot of risk can be mitigated through more frequent releases, and the cost is worth paying. You need to use judgment and decide for yourself.

Improve Efficiency to Increase Delivery Cadence

To follow the example, we have determined it takes ten people three days to release the code. From this we have derived that a monthly release is acceptable. However, several folks believe that with improved code quality, improved configuration management, better tooling to handle data migration, and regular rehearsals of the deployment procedures, it will be possible to cut the three-day time down to eight hours. Suddenly a release every two weeks looks viable. Perhaps every week is possible? What should you do?

My advice is to choose conservatively initially. Agree to a release every month. Let the organization prove that it can achieve this level of consistency. After a few months, reflect upon the code quality and instigate a program to improve configuration management. If slack resources are available, engage them to create tooling to improve data migration across schemas during a release. And finally, encourage the team to rehearse releases in a staging environment. Perhaps you'll need to buy, install, and commission such an environment. All of this will take time.

Challenge the team and the immediate function managers who control and perform releases to reduce the transaction and coordination costs. As these costs come down, review the progress at the operations review meeting and liaise with other stakeholders. When you feel confident that you can commit to a more frequent delivery cadence, such as bi-weekly, do so!

Reducing coordination and transaction costs is at the heart of Lean. It is waste elimination in its most potent form. It allows smaller batches to become efficient. It enables business agility. Reducing coordination and transaction costs is game changing. However, do not simply focus on reducing them. Reduce them with a goal in mind—to make more frequent deliveries of working software and thereby deliver more value to your customers more often.

Making On-Demand or Ad Hoc Deliveries

Regular delivery has advantages. Making a promise to deliver on specific dates, for example, every second Wednesday, allows those involved to schedule around it. It provides certainty. It can also reduce coordination costs because there is no overhead taken up in deciding when to make a delivery and who needs to participate—all of that is established once and is fairly consistent from then on.

Regular delivery also helps to build trust. Lack of predictability destroys trust. Failure to make a delivery on a promised date gets noticed much more than the specific content of a given delivery does.

Having made a strong case for a regular delivery cadence, it does make sense to have on-demand or ad hoc deliveries in some circumstances. What might those circumstances be?

First, on-demand or ad hoc delivery makes sense when the coordination costs of the release activity are small. When coordination costs are low, there is no benefit from scheduling coordination activities regularly. Secondly, it makes sense when the transaction costs are low, perhaps because the deployment of the code is largely automated and the quality is assured in advance, prior to deployment. And finally, it makes sense in environments where the deployments are so frequent that there is no real need to develop a pattern: New software is being delivered so often as to appear continuous to most observers and external stakeholders—their brains haven't been programmed to anticipate a delivery date. And when there is no expectation, there can be no disappointment.

This type of near-continuous deployment of code seems to be useful and necessary in some industries. The examples that have emerged from early Kanban adopters have been mostly in the media industry, for example, IPC Media in London, where they use multiple Kanban systems to plan development for online media properties such as mousebreaker.com, a highly addictive online game.

The first two circumstances of low coordination and low transaction costs would tend to indicate higher maturity organizations. This has also been observed with early adopters of Kanban. The Microsoft XIT department was partnered with a CMMI-ML5 vendor in India and Microsoft IT in Redmond, Washington, which is approximately CMMI-ML3 in maturity. High maturity organizations tend to have established levels of trust with value-chain partners and external stakeholders, including senior management, so they don't need a regular delivery cadence in order to build that trust.

So, in general, choose a regular delivery cadence except in circumstances in which trust already exists, high levels of capability and maturity already exist, and in domains where near-continuous deployment is desirable.

There is one final circumstance in which it is acceptable to make an on-demand delivery. That is when there is an urgent request that is being treated as a special case and expedited. This concept of an Expedite class of service is explained in chapter 11. We might choose to expedite for several reasons, the first and most obvious of which is in the event of a critical production defect. In circumstances where nothing else matters but to fix the problem, an off-cycle release should be planned.

There are other circumstances when an off-cycle release might make sense. Perhaps the sales team has taken an order from a big customer who wants a customized version of the software and because of budget restraints and the fiscal cycle, they need to take delivery before the end of the month (and the quarter). The order comes through from the operations group that software engineering has to drop everything to fulfill this one customer order because it is worth a great deal in revenue.

Under these circumstances, it makes sense to plan a special, off-cycle release. This release should be treated as exceptional, and the regular release cadence should be re-established as soon as possible following the exceptional release. It does pay to use some common sense, though. For example, if a regular release is scheduled for Wednesday and the exceptional release is required on Friday of the same week, it may make sense to delay the Wednesday release until Friday. If you do choose to do that, it is important to communicate it properly and sufficiently early so that expectations are reset. You don't want to lose trust with your value-chain partners as a side-effect of trying to be accommodating and helpful.

Takeaways

❖ "Delivery cadence" means an agreed-upon, regular interval between deliveries of working software.

❖ Kanban decouples delivery cadence from both development lead time and prioritization cadence.

❖ Short, time-boxed iterations have led to dysfunction with some teams attempting to adopt Agile development methods.

❖ Delivery or release of software involves coordination of many people from various functions. All of this coordination has a measurable cost.

❖ Delivery or release of software carries with it a set of transaction costs in both time and money. These costs can be determined and tracked.

❖ Efficiency of delivery can be calculated by comparing the sum of the transaction and coordination costs of making a delivery against the total cost (or burn rate) of creating the software for that delivery.

❖ Delivery cadence can be established by comparing the cost of producing a release against the value received from that release.

❖ Efficiency and cadence can be increased by focusing on reducing the transaction and coordination costs.

❖ Regular delivery builds trust.

❖ Setting an expectation of regular delivery and then delivering consistently against that expectation can be addictive.

❖ Scheduling regular deliveries reduces coordination costs.

❖ Ad hoc or on-demand delivery can make sense for high-maturity organizations with established, high levels of trust and low transaction- and coordination costs of delivery.

❖ Legitimate requests for expedited delivery may also be cause for an off-cycle release. Regular releases should be re-established as soon as possible after such a special, exceptional release.

❖   **Chapter 9**   ❖

Establishing an
Input Cadence

This chapter discusses the elements involved in agreeing on a suitable prioritization cadence and when or if it would make sense to have on-demand or ad hoc prioritization rather than a regularly scheduled prioritization meeting.

Coordination Costs of Prioritization

When we were introducing Kanban at Corbis, in 2006, we chose to start with the sustaining engineering effort that handled minor upgrade requests and production bug fixes for the full suite of IT applications, including functions such as finance and human resources as well as the more business-specific systems of the digital asset-management system and the e-commerce web site. These systems served at least six business units, including sales, marketing, sales operations, finance, and the function that supported the supply chain that sold digital photography, meta-data tagging, cataloging, and fulfillment—essentially the business's supply chain.

Six departments competed for the shared resources available to make these small changes and upgrades. When the kanban system was first introduced, a business case had been made for a sustaining engineering function that could provide frequent, tactical releases. This sustainment function would enable limited business agility through release of minor, incremental functions while other new IT application projects were built using a traditional program management office (PMO) governance mechanism. Each project in the portfolio was justified and authorized based on its own independent business case. The sustaining function had been approved by the executive committee, and an additional ten percent funding authorized for the software engineering function, which translated to an additional five headcount for the department. This new capability was named the Rapid Response Team (or RRT). The name was a complete misnomer—initially it was neither rapid, nor responsive, nor was there a team.

It wasn't feasible to create a specialist maintenance department with those five new people. Corbis had a considerable diversity of IT systems and many required specialized skills. The analysis function that developed and elaborated system requirements relied particularly heavily on specialists. The additional five heads were spread somewhat evenly across the software engineering function that included project management, system analysis, development, test, configuration management, and build engineering functions. So there was no team as such. The T in RRT was meaningless. The challenge for management was to show that this additional ten percent for resources was being spent on maintenance and sustaining work and wasn't simply being absorbed into major-project portfolio work.

It was decided to dedicate a project manager to the sustaining engineering effort. While this woman was not full-time on sustaining engineering, she did provide a single focal point for communication and coordination and counted as one-half of one of the five headcount allocated to the initiative. A build engineer from the configuration management team was also designated specifically to the initiative. His duties were to maintain the pre-production systems required for testing and staging, and to build code and push it into the test environment when required.

In order to maintain the integrity of the shared test environment, which was being used by multiple projects all at once, Corbis had a policy that only build engineers were permitted to promote code from the development environment to the test environment. This would change later, but in September 2006 the reality was that a build engineer was required to promote code for testing.

Prior to introducing Kanban, the coordination effort required to agree on a release scope for a maintenance release was prohibitive. The project manager, and often her boss, the group project manager, would convene a meeting of all the relevant parties, including business analysts, business representatives, system analysts, development managers, test leads, and the build engineer, and sometimes the configuration-management manager, as well as system-operations and help-desk personnel. These meetings could take several hours and were often inconclusive. Team members would be sent off to do estimates, and another meeting scheduled. The later meeting would often become bogged down in debate over priorities, and again, was often inconclusive. In September 2006 it was taking two weeks of calendar time with several meetings lasting many hours to agree on the scope for a release that was supposed to take only two weeks to build and deploy. Because of the two-week iteration length, only very small requests could be accommodated and many potentially valuable requests were ignored. These requests would have to be rerouted to a major project and so would likely take months or years before being implemented. The system was neither rapid nor responsive prior to the introduction of a kanban system. So the RR in RRT was also meaningless.

Kanban freed this team from all these dysfunctions and rightly earned the initiative its Rapid Response Team name.

When introducing the kanban system, the business owners were educated on the workflow, the input queue, and the pull mechanism. They learned that they would be asked simply to replenish empty slots in the queue and that it would not be necessary to prioritize the backlog of requests. If there were two slots free in the queue, the question would be, "Which two new items would you like next?" Assuming we had data on the average lead time and the due-date performance versus the lead-time target in the service-level agreement, the question might more elaborately become, "Which two items would you most like delivered 30 days from now?" The challenge then would be for six competing business owners to somehow agree to choose two items from the many possible choices.

Nevertheless, the question is a simple one, and it was suggested that answering such a simple question should easily be achievable in one hour. There was a consensus that one hour was reasonable, hence the business owners were asked if they would give up one hour of their week to attend a weekly prioritization meeting to refill the input queue for the sustaining engineering function.

Agreeing on a Prioritization Cadence

The meeting was scheduled for every Monday at 10 A.M. It was generally attended by higher-ranking business managers than used to come to the release scope meetings, usually the vice president of each functional group. In addition, the project manager, the Senior Director for Software Engineering, the Senior Director for IT Services (whose responsibilities included project management), at least one development manager, the test manager, the analysis team manager, and occasionally some additional individual contributors would attend.

Agreeing on a regular cadence provided everyone with predictability. They were able to set aside the one hour on Monday mornings, and generally the meeting was well attended.

Weekly is a good schedule for prioritization cadence. It provides frequent interaction with business owners. It builds trust through the interaction involved. In the collaborative, cooperative game of software development, it enables the players to move the pieces once a week. Weekly meetings are possible due to the simplicity of the question to be answered and the guarantee that the meeting can be completed in one hour. When business people take time away from running their businesses, they need to see that their time is well spent.

Many of kanban's aspects contribute to making the weekly prioritization meeting gratifying: It's a collaborative experience; there is transparency to the work and to the work flow; progress can be reported every week; and everyone gets to feel that they are contributing to something valuable. To several of the vice presidents at Corbis, it felt like the RRT process was making a difference. They gained a new level of respect for the IT department and they learned to collaborate with their peers in other departments in a way that previously hadn't been common at Corbis.

Efficiency of Prioritization

Weekly coordination meetings may not be the right answer for your organization. You may find that your coordination challenges are harder or simpler than those at Corbis. Some teams all sit together, so there is no need for a meeting; prioritization coordination amounts to a quick discussion across the desk. On the other hand, some teams may have people in multiple time zones and on several different continents, so weekly meetings may not be so easily scheduled. Perhaps the question to be answered won't be as simple as in the Corbis example and the meeting will take longer. It's hard to imagine a situation with more than six groups competing for the same shared resource, but it is possible. The more groups involved, the longer the meeting is likely to take. The longer the meeting, the less frequently you are likely to hold it.

As general advice, more frequent prioritization is desirable. It allows the input queue to be smaller, and, as a result, there is less waste in the system. WIP is lower and therefore the lead time is shorter. More frequent prioritization allows all parties to work together more often. The collaborative working experience builds trust and improves the culture. Strive to find the smallest, most efficient coordination scheme possible and hold prioritization meetings as often as is reasonable.

Transaction Costs of Prioritization

In order to facilitate an efficient meeting every Monday, Diana Kolomiyets, the project manager, would generally send an email on Thursday or Friday to inform attendees of the estimated number of free slots anticipated to be in the queue by Monday morning. She asked them to browse through the backlog of requests and pick out likely candidates for selection on Monday. This "homework" often prompted them to prepare some argument to support the successful selection of their favorite item. We began to see supporting documents at the Monday meeting. Some people would prepare a business case, some a presentation to support their choice. Others began to lobby each other. It was likely that someone on the prioritization board would take another to lunch on Friday, specifically to garner support for their choice at the Monday meeting. Horse-trading was introduced, so that one board member might agree to support another's item this week in exchange for support of their own item in a future week. The nature of the rules of the game, where multiple organizations competed for the shared RRT resource, introduced a whole new level of collaboration.

Occasionally, the business might be concerned that a request was too costly to implement in comparison to its worth, so they might solicit the analysis team to make an estimate. Later, rules regarding class of service were introduced to guide whether it was worthwhile to estimate an item or not. This is fully explained in chapter 11.

All of these activities, including estimation, business plan preparation, and candidate selection from the backlog, are preparatory work for prioritization. In economic terms these are the transaction costs of prioritization. It is desirable to keep these costs low. If the transaction costs become onerous, regardless of how low the coordination costs are, the team will not want to meet regularly. By avoiding detailed estimates as much as possible, the transaction costs are minimized, which facilitates more frequent prioritization meetings.

Improve Efficiency to Increase Prioritization Cadence

In general, the management team must be aware of all of the transaction and coordination costs incurred by everyone—not just the development team—involved in prioritization and selection of new items to queue for development and delivery.

Many Agile organizations use a form of prioritization called Planning Poker that uses a "wisdom of the crowds" technique, in which every team member gets to vote using a card representing a sizing number. The votes are averaged, or sometimes a consensus is sought by discussing outliers in the voting and then re-voting until everyone on the team agrees on an estimate. The poker cards often use a non-linear numbering scale, such as the Fibonacci Series, to encourage the idea of relative sizing.

Some argue that this planning technique, which is also a form of a collaborative cooperative game, is highly efficient, as it allows a fairly accurate estimate to be established quickly. There is anecdotal evidence to support this, but equally there is evidence to suggest that group-think is also possible. I've heard reports of teams, such as a startup in San Francisco, that consistently underestimated despite using a transparent collaborative game such as Planning Poker. I've also heard from senior managers at a well-known travel booking web site that teams consistently over-estimated, despite using Planning Poker. Whether or not you believe that these planning games are effective, the argument that they are efficient is worth considering more deeply.

It is true that planning games involving the whole team can create an estimate for an individual item, such as a user story, very quickly. However, the exercise involves the whole team. There is a significant coordination cost to this. It will work effectively on small teams focused on single products; however, if we extrapolate the technique to an organization like Corbis, where we were doing maintenance on 27 IT systems using 55 people, many of whom are specialists in one field, domain, system, or technology, it would be necessary to have almost all 55 people in the meeting to make a good estimate and achieve a "wisdom from the crowd." The transaction costs of planning and estimation may be small, but the coordination costs are high.

In general, because of this coordination-cost effect, these Agile planning methods are efficient only for small teams focused on single systems and product lines.

By choosing to eliminate estimation for most classes of service, both the transaction costs and the coordination costs of prioritization are reduced. This reduction facilitates much more frequent prioritization meetings because the meetings remain efficient. This has enabled kanban teams to make ad hoc or on-demand prioritization.

Making On-Demand or Ad Hoc Prioritization

As described in chapter 4, in 2004, Dragos Dumitriu introduced a kanban system with his XIT Sustaining Engineering team at Microsoft. The upstream business partners were four product managers who represented several business units. They focused and prioritized change requests for the 80 or so IT systems supported by XIT.

When Dragos and I designed the kanban system for introduction with XIT, we designed an input queue big enough to cope with at least one week of throughput. Despite the fact that all four business representatives and Dragos were based in Redmond, Washington, on the Microsoft campus, the prioritization meeting would take place by phone. The Microsoft campus is huge. Buildings numbers go into the hundreds, although there are actually only about 40 buildings in total. The area covered is several square miles and transport between sections of the campus is by minibus or Toyota Prius. Many "softies" prefer conference calls for coordination meetings rather than face-to-face. This has a negative impact on the level of trust and social capital in their workforce, but it facilitates efficiency.

So Dragos established a weekly phone call to prioritize new change requests into their backlog. The four product managers represented business units that provided funding via inter-company budget transfer to fund Dragos' team. Based on this funding, it was possible to determine roughly how many times someone should get to pick an item from the backlog. The product manager who provided six tenths of the funding would get to pick three from every five opportunities. Others would get to select items in a similar way based on their level of funding. The product manager who provided least funding would get to pick roughly once in every 11 times. We might describe this as a weighted round-robin method of selection.

So the rules of the XIT prioritization collaborative cooperative game were simple. Each week the product managers would refill the open slots in the input queue—typically three slots. Each of them would get to choose based on their position in the round-robin queue. The target lead time in the service-level agreement was 25 days. So if they got a chance to choose a change request for development, they would ask themselves, "Which of the items in my backlog do I most want delivered 25 days from now?" The order in which they got to choose was very clear and simple based on their level of funding for the department.

Due to the simple nature of these rules, the meeting was finished very quickly. It became clear that a coordinating phone call really wasn't necessary. Dragos had the Microsoft Product Studio (a forerunner to Visual Studio Team System, Team Foundation Server) database provide an email from a trigger that indicated when a slot became free. He would then forward that email to the four product managers. They would quickly agree whose turn it was to choose an item and that person would select something. Typically, an empty slot in the queue was replenished within two hours.

The exceptionally low coordination costs, coupled with the low transaction costs related to the decision not to estimate change requests, along with the relatively high maturity of the team involved, allowed Microsoft XIT to dispense with regularly scheduled prioritization meetings.

It is worth noting that Microsoft in Redmond is roughly the equivalent of a CMMI-ML3 organization and that the vendor being used for XIT development and testing was a CMMI-ML5 team based in Hyderabad, India. So this team had the advantage of low coordination costs, low transaction costs, and particularly high levels of organizational maturity. The net effect of all three meant that on-demand prioritization meetings made the team more effective.

As a general rule, you should choose ad hoc or on-demand prioritization when you have a relatively high level of organizational maturity, low transaction costs, and low coordination costs of prioritization. Otherwise, it is better to use a regularly scheduled prioritization meeting and coordinate selection of input queue items with a regular cadence.

Takeaways

❖ "Prioritization cadence" means an agreed-upon regular interval between meetings to prioritize new work into the input queue for development.

❖ Kanban removes potential dysfunction around the coordination of iteration planning in Agile methods by decoupling the prioritization cadence from the development lead time and delivery.

❖ Prioritization of new work requests such as user stories involves coordination of many people from various functions. All of this coordination has a measurable cost.

❖ Estimation to facilitate prioritization decisions represents the transaction costs in both time and money associated with prioritization. These costs can be determined and tracked.

❖ Policies concerning the method of prioritization and the inputs for decision making represent the rules of the collaborative cooperative game of prioritizing in Kanban applied to software development.

❖ Planning games used in Agile methods do not scale easily and can represent a significant coordination cost for larger teams with broader focus than a single product line.

❖ Prioritization cadence can be established by encouraging those involved in prioritization decision making to meet as regularly as is reasonable based on the transaction and coordination costs involved.

❖ Efficiency and cadence of prioritization can be increased by focusing on reducing its transaction and coordination costs.

❖ Frequent prioritization meetings build trust.

❖ Scheduling regular prioritization meetings reduces coordination costs and is particularly useful in lower-maturity organizations.

❖ Ad hoc or on-demand prioritization can make sense for high-maturity organizations with established high levels of trust and with low transaction and coordination costs associated with the policies for prioritization decision making.

❖　Chapter 10　❖

Setting Work-in-Progress Limits

As discussed in chapter 2, one of the five core properties in the Kanban Method is that work-in-progress is limited. So it's true to say that one of the most important decisions you'll make when introducing Kanban is choosing limits for work-in-progress throughout the workflow.

Chapter 15 advises that the work-in-progress limits should be agreed upon by consensus with up- and downstream stakeholders and senior management. It is true that limits could be unilaterally declared; however, there is power in gaining a consensus and obtaining a commitment from external stakeholders regarding the WIP-limit policy. When your team and process is put under stress you can fall back on the collaborative agreement around which there is consensus. You can redirect the discussion to a redefinition of the process rather than agree to bend, stretch, or otherwise misuse the system as designed and implemented. Building consensus is a way to maintain WIP-limit discipline and avoid having to override or abandon a limit.

Limits for Work Tasks

At Microsoft with the XIT team, Dragos Dumitriu decided that developers and testers should work on a single item at a time. There would be no multi-tasking. This was unilaterally declared, but fortunately this choice did not prove problematic with other stakeholders. This was in line with current working practices and the Personal Software Process (PSP) method in use with the team. The organization was mature enough to maintain discipline and follow the process that had been agreed upon. You may recall that at the beginning of the case study, in the fall of 2004, there were three developers and three testers on the team. So the WIP limit for each, development and test, was three.

At Corbis in 2006, with the sustaining engineering initiative, we made a similar decision; that analysts, developers, and testers should generally work on only one customer-valued work item at a time. With new major projects, we tended to make different decisions. There was more collaborative work on those projects. It was common for teams of two or three people to work on a single item. Because those items could become blocked or delayed, we speculated it might make sense to allow some task switching and some additional parallelism; hence the WIP limit was set to infer two or three people per item, but to allow some overflow. For example, if we had ten people and anticipated two people per item, the WIP limit might be five plus a few more to smooth the impact of a blockage. Perhaps eight (five plus three) would be the right limit in such circumstances.

There has been some research and empirical observation to suggest that two items in progress per knowledge worker is optimal. This result is often quoted to justify multi-tasking. However, I believe that this research tends to reflect the working reality in the organizations observed. There are a lot of impediments and reasons for work to become delayed. The research does not report the organizational maturity of the organizations studied, nor does it correlate the data with any of the external issues (assignable-cause variations, discussed in chapter 19) occurring. Hence, the result may be a consequence of the environments studied and not indeed an ideal number. Nevertheless, you may encounter resistance to the notion that one item per person, pair, or small team is the correct choice. The argument may be made that such a policy is too restricting. In that case, setting a WIP limit of two items per person, pair, or team is reasonable. There may even be cases where a limit of three per person, pair, or team is acceptable.

There is no magic formula for your choice. It is important to remember that the number can be adjusted empirically. You can select a number and then observe whether it is working well. If not, adjust it up or down.

Limits for Queues

When work is completed and waiting to be pulled by the next stage in your workflow, it is said to be "queuing." How big should these queues be? As small as possible. The

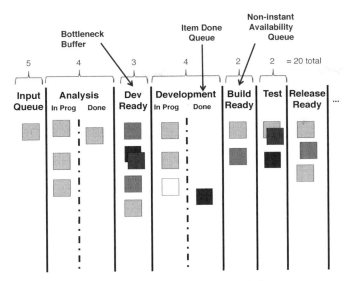

Figure 10.1 Card wall showing different types of queues and a buffer

WIP limit for a queue is often bracketed with its preceding work step. For example, the Development and Development Done queues are bracketed together, as shown in Figure 10.1. If a really tight policy on work task WIP was established, such as strictly one item per person, pair, or small team, it will be necessary to have some queue to absorb variation and maintain flow. If, in operation, your kanban system suffers from stop-go behavior that causes workers to be idle due to variability in the length of time it takes to complete tasks, you may need to increase the queue sizes. However, if you already made a choice to have, for example, two items in progress per person, pair, or team, you already have buffered for variability, so your queue size can be effectively zero. Simply bracket the work task column and the complete queue together.

Buffer Bottlenecks

The bottleneck in your workflow may require a buffer in front of it, as shown in Figure 10.1. This is a typical bottleneck exploitation mechanism, as explained in chapter 16. The sizing of the buffer is important. Again, you want it to be as small as possible. Buffers and queues add WIP to your system and their effect is to lengthen lead time. However, buffers and queues smooth flow and improve the predictability of that lead time. By smoothing flow, they increase throughput, so more work is delivered through the kanban system. Buffers also ensure that people are kept working and provide for greater utilization. There needs to be balance, and buffers help maintain it. In many instances you are seeking business agility through shorter lead times, and higher quality partly through lower work-in-progress. However, do not sacrifice predictability in order to achieve agility or quality. If your queue and buffer sizes are too small and your system suffers from a lot of stop-go behavior due to variability, your lead times will be unpredictable, with a wide spread of variability. The key to choosing a WIP limit for a buffer is that it must be large enough to ensure smooth flow in the system and avoid idle time in the bottleneck. More detail on buffer sizing and how to design buffers for both capacity-constrained and non-instant-availability bottlenecks is discussed in chapter 16.

Input Queue Size

The size of the input queue can be directly determined from the prioritization cadence and the throughput, or production rate, in the system. For example, if a team is producing at a mean rate of five completed work items per week (with a typical range of four to seven items per week), and the queue replenishment cadence is weekly, the queue size should probably be set to seven. Again, this can be empirically adjusted. If you run your system for several months and the queue is never totally depleted before your prioritization meeting occurs, it's probably too large, so reduce it by one and observe the results. Repeat until you have a prioritization meeting in which you are asking the business representatives to refill the entire queue.

If, on the other hand, you have a weekly prioritization meeting on a Monday and the queue is depleted by Thursday afternoon—and some of the team was idle as a result—your queue is too small. Increase the queue size by one and observe for a few more weeks.

Queue and buffer sizes should be adjusted empirically as required. Hence, do not fret over a decision to establish a WIP limit. Do not delay rollout of your kanban system because you can't agree on the perfect WIP-limit numbers. Choose something! Choose to make progress with imperfect information and then observe and adjust. Kanban is an empirical process.

What size should the input queue be if you are using on-demand prioritization? You may recall from chapter 4 that the XIT team had an input queue of five items. This was designed to be large enough to absorb one week's throughput. It was based on the assumption that the prioritization meeting was happening weekly. However, very quickly the product managers decided that the meeting was unnecessary and that it was acceptable to make event-driven decisions when a slot in the queue became free. Once this happened, I should have advised Dragos to reduce the input from five to only one. It's a reflection of my inexperience at the time that I didn't do this. The system had changed. The assumptions on which it was designed had changed. The input queue size policy was based on those assumptions and should have been revisited. Had we done so, the lead-time improvements would have been even more impressive.

When XIT switched to on-demand prioritization it typically took them two hours to refill an empty slot in the queue. It would have been acceptable to assume that the longest time to reset the queue would have been four hours. However, the developers were not collocated with the product managers. The prioritization decision makers were in Redmond, Washington, and the developers in Hyderabad. Each of them in turn was working (officially) eight-hour days at opposite ends of the clock. So it is likely that there might have been occasions when the Indians turned up for work in the morning, finished off a task, and needed their queue replenished, but the product managers are oblivious, asleep in bed. Given this non-instant availability problem, we probably should have allowed 16 hours to replenish a single item in the queue under extreme circumstances. Remember that the developers were the bottleneck in this workflow. In order to maximize throughput, we never want those developers to be idle. So we need to be conservative; 16 hours is conservative when the average queue replenishment decision takes only two hours. So what would the throughput be in an average 16 hour period? At peak performance, the team achieved 56 items in a single quarter. That's less than five per week. So in a 16-hour period it is unlikely they will complete even a single item. So a queue size of one is perfectly acceptable. No queue at all would be unacceptable. There is still some chance that the team would suffer idle time when they finish an item during the 16 hour window when the product managers may be unavailable to refill the queue.

Unlimited Sections of Workflow

In the Theory of Constraints pull-system solution for flow problems, known as Drum-Buffer-Rope, all work stations downstream from the bottleneck have unlimited WIP. This design is based on the assumption that they have a capability of greater throughput than the bottleneck and have slack capacity that results in idle time. As a result there is no need for a WIP limit. This is illustrated in Figure 10.2(a) which is based on the metaphor used in Goldrat's *The Goal* and shows a patrol of scouts hiking in single file. A rope is attached between the lead scout and the slowest moving (no. 4 in line), who is the bottleneck on throughput (the rate the patrol covers ground). Only one rope is needed as the scouts behind the slowest, will never fall behind, as they can walk faster than the fourth in line who constrains the pace of the whole patrol.

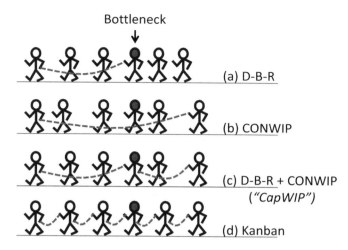

Figure 10.2 Stick men illustrating four different WIP-limited pull system designs

With a kanban system, most or all stations in the workflow have WIP limits. This has a potential advantage because impediments caused by unexpected, unanticipated variability may cause an upstream step to become a temporary bottleneck. The local WIP limit with the kanban system will stop the line quickly, keeping the system from clogging and becoming overloaded. When the impediment is removed, the system will then restart gracefully. Kanban-style WIP limiting is illustrated in figure 10.2(d), which shows how the scouts would be roped together if using a kanban style. In this case, each scout is roped to the next in line. Five ropes are required to control the pace of the entire patrol of six scouts.

In some cases, it may be acceptable for a kanban system to have unlimited downstream process steps. In the XIT example at Microsoft, it was assumed that the user base available to perform acceptance testing was effectively infinite and essentially instantly available, hence there was no need to limit WIP in user acceptance testing. At Corbis, the release-ready queue was unlimited. This was based on the assumption that the batch of release-ready work would never become excessive, given the agreed upon bi-weekly release cadence. If, on the other hand, it had been possible for release-ready material to become excessive, that would have raised the complexity of the release, which would have resulted in the coordination effort and the transaction costs of the release becoming uneconomical, it would have been necessary to limit WIP in the release-ready queue. However, this was never the case at Corbis, so the release-ready state remained unrestricted.

Don't Stress Your Organization

Choosing overly tight WIP limits initially may put excessive stress on your organization. Lower-maturity organizations with poorer capabilities will have more impediments. Hence, lower-maturity organizations with poorer capabilities may find that introducing a kanban system causes excessive pain if WIP limits are set too low. If there are a lot of impediments, represented by lots of pink tickets across the card wall, overly tight WIP limits will mean that everything grinds to a halt and a lot of people will be idle. While idle time tends to focus attention and accelerate efforts to resolve issues and remove impediments, it may be just too painful for a lower-maturity organization. Senior managers can become irritable observing too many idle people still collecting paychecks.

When introducing change, you need to be aware of the J-curve effect. Ideally, each change produces a little j where any impact on performance is shallow, then the system quickly recovers and shows improvement. If you make the WIP limits too tight, you will suffer a J-curve effect that is too deep and too long, which may cause undesirable effects: Kanban is exposing all the problems in the organization, but it may end up being blamed for making everything worse, and will be seen as part of the problem rather than the solution. So tread carefully. With more capable, more mature organizations that suffer few unexpected issues (assignable-cause variations) you can be more aggressive with your WIP-limit policies. For more chaotic organizations, you will want to introduce looser limits initially with greater WIP and an intention to reduce it later.

It's a Mistake Not to Set a WIP Limit

Although I caution you not to be aggressive when setting initial WIP limits, I have become convinced that not setting WIP limits is a mistake.

Some early adopters of Kanban, such as Yahoo!, chose not to set WIP limits because they assumed their teams were too chaotic to cope with the pain that it can introduce. The hope was that these organizations would mature through the visual-control elements of Kanban and that WIP limits could be introduced later. This proved problematic, however, and several teams abandoned Kanban without seeing much improvement, while others were disbanded in corporate reorganizations or project cancellations, denying us further data points. At Corbis, several teams on major projects pursued Kanban with only very loose WIP limits on course-grained, high-level functionality. The results were somewhat mixed.

I've become convinced that the tension created by imposing a WIP limit across the value-stream is positive tension. This positive tension forces discussion about the organization's issues and dysfunctions. The dysfunctions generate impediments to flow and result in sub-optimal productivity, lead time, and quality. The discussion and collaboration invoked by the positive tension of a WIP limit is healthy. It is the mechanism that enables the emergence of a continuous-improvement culture. Without WIP limits, progress on process improvement is slow. Teams that have imposed WIP limits from the beginning have reported accelerated growth in capability and organizational maturity and have delivered superior business results with frequent, predictable deliveries of high quality software. In comparison, teams who have deferred introduction of WIP limits have generally struggled and showed only limited improvement.

Capacity Allocation

Once we establish WIP limits for the flow through the system, we can consider capacity allocation by work item type or class of service.

Figure 10.3 shows the card wall design from chapter 6 with WIP limits across the columns totaling 20 cards. The capacity is allocated across work item types, namely, 60 percent for change requests, 10 percent for maintenance items, and 30 percent for

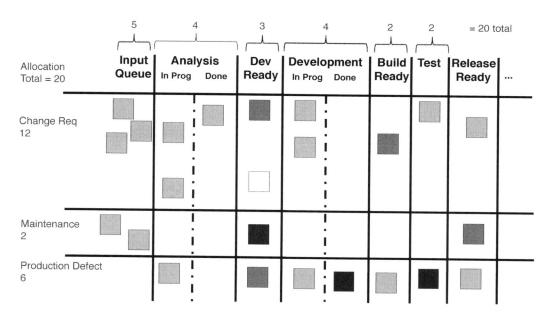

Figure 10.3 Card wall showing swim lanes for each work item type with explicit WIP limits for each lane

production text changes. This equates to a swim-lane WIP limit of 12 for change requests, 2 for maintenance items, and 6 for production text changes.

Capacity allocation allows us to guarantee service for each type of work received by the kanban system. The allocation should generally be made in response to the comparative demand observed for each type of work. Hence, it is important to complete some demand analysis to facilitate reasonable allocation of WIP limits on swim lanes for each type of work.

Takeaways

❖ WIP limits should be agreed upon through consensus with up- and down-stream stakeholders and senior function management.

❖ Unilateral declaration of WIP limits is possible but may prove difficult to defend later, when the system is placed under stress.

❖ WIP limits for work tasks should be set as an average number of items per person, developer pair, or small, collaborative team.

❖ Typically, the limit should be in the range of one to three items per person, pair, or team.

❖ Queue limits should be kept small, typically only large enough to absorb the natural (random-cause) variation in size of items and task duration.

❖ Bottlenecks should be buffered.

❖ Buffer sizes should be as small as possible, but large enough to ensure optimal performance in the bottleneck and maintenance of flow in the system.

❖ All WIP limits can be adjusted empirically.

❖ Kanban is an empirical process.

❖ Excessive time should not be wasted trying to determine the perfect WIP limit; simply pick a number that is close enough, and make progress. Empirically adjust if necessary.

❖ Unlimited downstream sections of workflow are possible.

❖ Care should be taken that establishing unlimited workflow steps does not introduce bottlenecks or cause excessive transaction- or coordination costs when deliveries (batch transfers downstream) are made.

❖ Once WIP limits have been set, capacity can be allocated across work item types.

❖ Swim lanes are often used for each work item type and a WIP limit set for each swim lane.

❖ Capacity allocation requires comparative demand analysis across the different types of work received by the kanban system.

❖ **Chapter 11** ❖

Establishing Service Level Agreements

We are all familiar with the concept of differing classes of service. Anyone who's checked in for a flight at an airport understands that customers who pay more for their ticket, or who enjoy the rewards of a customer-loyalty program, are permitted to use an express lane to "cut in line." Sometimes these privileges extend to airport security lines and include the use of a special lounge and preferential treatment at boarding time. Customers who pay more or who spend money with the airline on a regular basis enjoy a better class of service.

We are familiar with this concept in software development and IT systems work, too, most evidently with defect resolution, and particularly with production defects. We assess defects by severity (impact) and priority (urgency). High severity, high priority defects are fixed immediately. They receive a different, higher, class of service than other work. To fix a high-severity production defect, we put other work aside, pull in as many people as we require, and often make special plans for an emergency fix, patch, or release to alleviate the problem.

This concept can be applied more generally, which offers some advantages in both business agility and risk management. Some requests are needed more quickly than others, while some are more valuable than others. By offering to treat different types of work with different classes of service, we can offer the customer more flexibility while optimizing the economic outcome.

Classes of service provide us a convenient way of classifying work to provide acceptable levels of customer satisfaction at an economically optimal cost. By quickly identifying the class of service for an item, we are spared the need to make a detailed estimate or analysis. Policies associated with a class of service affect how items are pulled through the system. Class of service determines priority within the system. Classes of service allow for a self-organizing, value- and risk-optimized approach to prioritization and "re-planning."

Typical Class-of-Service Definitions

Classes of service are typically defined based on business impact. Different-colored sticky notes, index cards, or tickets are assigned to each class and clearly signify the class of service of any given request, as in Figure 11.1; otherwise, separate swim lanes are drawn on the card wall to signify membership in a class of service.

Each class of service comes with its own set of policies that affect how items are prioritized when they are pulled through the kanban system. The class of service also comes with an explicit promise to the customer. Following is a brief example of a set of class-of-service definitions. While this set isn't a precise facsimile of those used with any specific Kanban implementation, it does represent a strong generalization of classes of service observed in the field.

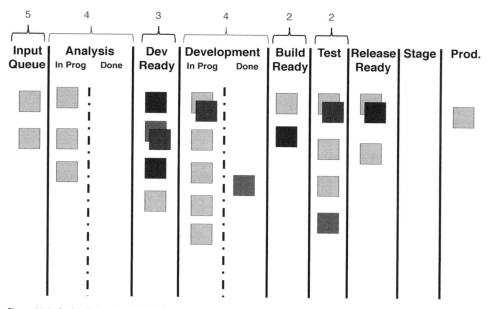

Figure 11.1 Card wall showing colored tickets to signify class of service

In this example set, four classes of service are offered. As a general guideline, you may want to offer up to a maximum of six such classes. Too many will become too complicated to administer and operate. The number of classes of service should be few enough so that everyone involved—team members and external stakeholders—can remember them all, while sufficient enough to offer flexibility in response to customer demand.

Expedite

The Expedite (or "Silver Bullet") class of service is well understood in the manufacturing industry. A typical scenario might be a sales team attempting to hit a quarterly sales target coupled to a customer who has budget that must be spent by the end of a fiscal year. The customer has been delaying a purchasing decision; he finally makes a choice just as time runs out on the current fiscal year. The order is placed, but it is conditional on delivery before the deadline. The manufacturer agrees to a price and quantity  and accepts the order. The order must be fulfilled, delivered, and invoiced before the final day in the quarter. Such an order typically hits the factory via the regional sales vice president's office with a request to expedite delivery given the tight timeframe and value of the order.

The ability to expedite offers a vendor the ability to say "Yes!" in difficult circumstances to meet a customer need. However, expediting orders badly affects manufacturing supply chains and distribution systems. Expediting is known in industrial engineering and operations research both to increase inventory levels and to increase lead times for other, non-expedited orders. The business makes a choice to realize value on a specific sale at the cost of both delaying other orders and incurring the additional carrying costs of higher inventory levels. If the company is well governed, the value generated from expediting will exceed the costs incurred from longer lead times (and potential lost business as a result) and the cost of carrying the higher inventory level.

Manufacturing companies often create policy to limit the number of expedite requests that hit a factory. One common policy is to grant a fixed number of so-called "silver bullets" to a regional sales vice president in a given time period. Hence, the term "silver bullet" has become synonymous with expediting in manufacturing or distribution.

Unfortunately for clarity's sake, the term "silver bullet" was already in use in software engineering. Fred Brooks defined it as a single change (in technology or process) that would create an order of magnitude (ten times) improvement in programmer productivity. Hence, I recommend that you stick with the term Expedite for this class of service. In companies that do manufacturing, or where senior management is familiar with manufacturing, however, I've observed that they prefer the term "silver bullet." This is fine so long as the technology people realize the difference in usage.

Fixed Delivery Date

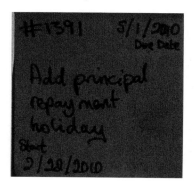

In mid-February 2007, a developer entered my office to ask if I was aware of an issue with a service platform we used for credit-card processing. I was not, so he explained. Apparently, the vendor had found that their codebase was too difficult to maintain as they sought to add more and more features to their platform—a common ailment in software development. In order to meet demands for new features in 2006, they had completely replaced their platform with a new system that had a completely new application programming interface (API). They had informed all of their customers and given them 15 months' notice that the old system would be switched off after March 31, 2007. Put another way, if we didn't upgrade our systems to use the new platform, we would cease trading on the Internet on April 1, 2007. This would be significantly inconvenient for a business that made much of its revenue through web-based sales, not to mention the embarrassment for the firm's owner. We had only six weeks to make the necessary changes and deploy the new code to production. The ticket for this work entered our kanban system marked with a fixed delivery date. The additional information on the ticket was intended to draw attention to the cost and to the impact of late delivery, as well as to enable the team to self-expedite the item and ensure on-time delivery.

It was not the first time we had seen such a request. An earlier request related to the integration of IT systems from an acquired company. The attached "fixed" date had been derived from the business case for the acquisition, which had shown significant cost savings from February 1 of that same year.

A theme, or pattern, seemed to be emerging: Some requests related to major contractual obligations, some to regulatory requirements (usually from the federal government), and some to strategic initiatives, such as the acquisition of another business.

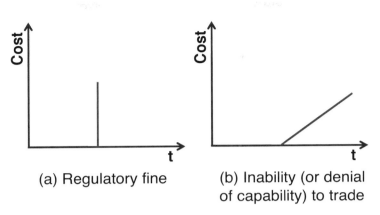

(a) Regulatory fine

(b) Inability (or denial of capability) to trade

Figure 11.2 Two cost of delay function profiles for Fixed Date class of service

Requests of this nature carried a significant cost of delay, whether direct or indirect, which tended to fit into one of two categories: There would be a date when a penalty (or a fine)—a direct, specific, out-of-pocket cost—would be incurred, imposed by the regulatory authority or the terms laid out in the contract. Alternatively, there would be a requirement to cease some activity, such as selling a particular type of item or operating in a particular jurisdiction, until the requirements were met. This second, indirect cost is a cost of lost opportunity—the potential lost revenue during the period of delay. Both types are charted in Figure 11.2.

Businesses with seasonal calendars, such as schools and colleges, tend to have hard calendar constraints. If you are working in a sector like education, you may have customers who want delivery of software at fixed times of year or not at all: Failure to deliver in their window results in a lost sale. Anything that has a physical or cultural "launch window" should be considered as having a (near) unit-step cost-of-delay function, and should be treated as having a fixed delivery date, if that future date falls within a reasonable lead-time window from the present time.

Standard Class

Most items needed with some urgency should be treated as standard class items. The policies and service-level agreement for a standard class item may vary by work item type. One common kanban system design scheme separates work types by size, such as small, medium, and large. A different service-level agreement for standard class items of each size can be offered. For example, small items are typically processed within four days, medium-sized items within one month, and large items within three months. Standard class items tend to have a tangible cost of delay that can be calculated (though not necessarily in monetary units). The cost of delay would tend to be incurred soon—within the timeframe for delivery of the request. The cost of delay tends to be immediate: If we had this function today, benefit would be derived tomorrow!

Intangible Class

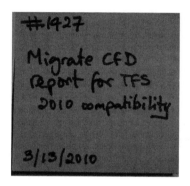

It makes sense to offer a fourth, lower class of service. I have struggled to find a suitable name for this class and have settled on "intangible." I'm not entirely happy with it, so it may be subject to change in a later edition of this book. Intangible class items may be important and valuable, but there is no tangible cost of delay associated with them in the near future. That is, there is no cost of delay within the timeframe that it might take to deliver the item. Requests that fit this pattern relate to deliverables with potentially fixed dates that are far in the future, such as platform replacement.

For example, in 2005 Microsoft launched SQL Server 2005, the latest version of its RDBMS database server. The 2005 edition replaced the 2000 edition, which then became "end of life." As the dominant player in the market, Microsoft is required to support its products for ten years after they are introduced. As a result, support for SQL Server 2000 would need to continue until 2010. This provides customers with a five-year window to replace code that is incompatible with the newer, 2005- (or 2010-) editions of the platform. So in 2005 or 2006, replacing database code—stored procedures, persistence code—is not a time-critical priority. There is no cost of delay incurred in those years. However, as time goes by and the code is not changed, the cost increases. It becomes increasingly difficult to work with other products, as newer versions require SQL Server 2005 as a prerequisite. More and more pressure builds up to make the switch to the newer platform. By 2009 the matter is urgent, as Microsoft will cease support for the older product, and failure to upgrade will leave the business running old machines with older, unsupported operating systems and associated infrastructure. If this risk is unacceptable, the code must be upgraded. This platform-replacement problem is a common one that software engineering teams struggle with continually. There is a desire to start the work early and complete it in a timely manner, but capacity to do the upgrade work is displaced by other work that is more urgent or critical. In other words, platform replacement, although it has a low immediate cost of delay, gets displaced by other work with a greater and more immediate cost of delay.

It does make sense to offer a class of service that would allow such work to be undertaken early. Capacity can be set aside for the work to ensure that it is completed; however, there may be no time guarantee. In addition, it is this low-cost-of-delay work that will always be set aside when more urgent requests come along. In order to have slack to process an expedite request, there must be low-cost-of-delay work that can be displaced while the expedite request is processed. These intangible class items provide that slack.

Policies for Class of Service

A visualization technique should be employed to readily identify the class of service. As mentioned earlier, either different colors of tickets or different swim lanes on the card wall are the most common. Some teams have used decorations such as stick-on stars attached to the work item ticket. A swim lane for expedite requests is also a common choice. How to visualize your classes of service is up to you. This chapter uses different colors to signify class of service. The goal is to ensure that any staff member, on any given day, can use the simple prioritization policies associated with a class of service to make good quality prioritization decisions in the field without management intervention or supervision.

Following is an example of prioritization policies for the four classes of service defined previously. Naturally, with every implementation, the class-of-service definitions will be unique and their policies for use will differ from these examples. These policies are based on empirical evidence and they fairly accurately reflect policies real teams have used.

Expedite Policies

- Expedite requests use white cards.

- Only one expedite request is permitted at any given time. In other words, the Expedite class of service has a WIP limit of 1.

- A qualified resource must pull Expedite requests immediately. Other work will be put on hold to process the expedite request.

- At any point in the workflow, the WIP limit may be exceeded in order to accommodate the expedite request. Capacity is not being held in reserve for expediting.

- If necessary, a special (off-cycle) release will be planned to put the expedite request in production as early as possible.

Fixed Delivery Date Policies

- Fixed delivery date items use purple cards.

- The required delivery date is displayed on the bottom right-hand corner of the card.

- Fixed delivery date items receive some analysis and an estimate of size and effort may be made to assess the flow time. If the item is large it may be broken up into smaller items; each smaller item will be assessed independently to see whether it qualifies as a fixed delivery date item.

- Fixed delivery date items are held in the backlog until they are selected for the input queue, close to the ideal point in time at which they can be delivered on time given the flow-time estimate.

- Fixed delivery date items are pulled in preference over other, less risky items. In this example, they are pulled before standard or intangible class items.

- Fixed delivery date items must adhere to the WIP limit.

- Fixed delivery date items queue for release when they are complete and ready for release. They are released in a regularly scheduled release just prior to their required delivery date.

- If a fixed delivery date item gets behind, and release on the desired date is at risk, its class of service may be promoted to an expedite request.

Standard Class Policies

- Standard class items use yellow cards.

- Standard class items are prioritized into the input queue based on an agreed-upon mechanism, such as democratic voting, and are typically selected based on their cost of delay or business value.

- Standard class items use first in, first out (FIFO) queuing as they are pulled through the system. Typically, when given an option, a team member pulls the oldest standard class item if there is no expedite or fixed date item to choose in preference.

- Standard class items queue for release when they are complete and ready for release. They are released in the next scheduled release.

- No estimation is performed to determine a level of effort or flow time.

- Standard class items may be analyzed for order of magnitude in size, typically, small (a few days), medium (a week or two), and large (perhaps months).Classes of service should be clearly, visually displayed by using, for example, different colored cards to represent the class of service or different swim lanes on the card wall.

- Large items may be broken down into smaller items. Each item may be queued and flowed separately.

- Standard class items are generally delivered within x days of selection with a due date performance of m percent.

A typical standard class service-level agreement might offer a 30-day lead time with 80 percent due-date performance. In other words, four out of five requests should be delivered within 30 days.

Intangible Class

- Intangible class items use green cards.

- Intangible class items are prioritized into the input queue based on an agreed-upon mechanism, such as democratic voting, and are typically selected based on some longer-term impact or cost of delay.

- Intangible class items are pulled through the system in an ad hoc fashion. Team members may choose to pull an intangible class item regardless of its entry date, so long as a higher-class item is not available.

- Intangible class items queue for release when they are complete and ready for release. They are released in the next scheduled release or are held to be assembled with other items.

- No estimation is performed to determine a level of effort or flow time.

- Intangible class items may be analyzed for size. Large items may be broken down into smaller items. Each item may be queued and flowed separately.

- Typically, an intangible class item is put aside in order to process an expedite request.

- It may not be necessary to offer a service-level agreement with intangible class items. If it is necessary, it should be a significantly looser agreement than that offered for standard class items, for example, 60 days with 50 percent due-date performance.

Determining a Service Delivery Target

In the example set of classes of service above, the Standard class of service used a target lead time, for example, 28 days (4 weeks). The concept of offering a target lead time coupled with a due-date performance metric is an alternative to treating each item individually and having to estimate and commit to a delivery date for each item. The service-level agreement allows us to avoid costly activities, such as estimation; low-trust activities, such as making commitments; and to spread risk by aggregating a large collection of requests and promising only aggregate performance in the form of a percentage due-date performance. By avoiding making promises we are unlikely to be able to keep, we avoid the danger of losing the trust of our customers. Therefore, it's important to communicate that the target lead time in the Standard class of service is just that, a target!

To determine the target lead time, it helps to have some historical data. If you don't have any, make a reasonable guess. If you do, then the most scientific means of determining the target lead time is to process the lead times (from first selection until delivery) through a statistical process-control package or kanban tracking tool that supports statistical process control (such as Silver Catalyst) and use the upper control limit (the plus-3 sigma limit) for your lead time. This ensures a time that you can hit under most normal circumstances and miss only when there is a genuine assignable-cause problem (see chapter 19 for a more detailed explanation).

If, on the other hand, that last paragraph meant nothing to you, then a more lay-person's explanation is that you want the lead time to be achievable most of the time, but also aggressive enough that it keeps the team focused. It is likely that your work items vary in size, complexity, risk, and expertise required. If so, the lead times vary significantly. That's okay. If you perform a spectral analysis of some historical data and can see that perhaps 70 percent are delivered within 28 days, and the remaining 30 percent spread out over another 100 days, then perhaps it's reasonable to suggest a target delivery date of 28 days.

I've learned that the use of classes of service is a very powerful technique. With my team in 2007, approximately 30 percent of all requests were late compared to the target lead time. We reported this as the Due Date Performance metric. It was never above 70 percent. However, despite this dismal performance versus the target date, we had very few complaints. The reasons for this became evident: All the important items—those with high risk or high value—were always on time, and there was a trust that the late ones would be delivered within an additional two or four weeks, as deliveries were happening with dependable regularity.

The Expedite and Fixed Delivery Date classes of service were ensuring that important items were always on time. Meanwhile, the other Standard class items that were late were generally delayed by only one or two releases (14 or 28 days, respectively). The customers trusted the release cadence. That trust was earned by doing. We consistently shipped a release every second Wednesday. With the insignificant cost of delay associated with many Standard (and Intangible, as they were not delineated separately) class items, the business focused on what had been delivered and planning for future items rather than worrying much about precise delivery dates for work-in-progress.

This result was significant because Kanban with classes of service had clearly changed the customer psychology and significantly changed the nature of the relationship and the expectations. The customers were now oriented around the long-term relationship and the performance of the system, and not on the delivery of any specific item or items. This gave the development team the freedom to focus on the right things and not waste time addressing issues that stemmed from a low level of trust between them and their customers.

Assigning a Class of Service

The class of service for an item should be assigned when the item is selected into the input queue. If an item is an expedite request, this should be self-evident—it comes with a request for the item to be processed as soon as possible. This is justified based on a business case that shows some immediate opportunity or identifies a significant cost that will be incurred if the request is not fulfilled. Perhaps that cost is already being incurred. This is typical with high-severity production defects.

If an item has a fixed delivery date this also should be self-evident by its nature. Perhaps the request relates to new regulatory requirements laid down by an independent regulatory authority, or to some seasonal nature of the business. If an item is of the fixed date class, typically that date is known, lies within a reasonable timeframe—perhaps twice as long as the typical standard class of service target lead time—and the item may have been estimated so as to be introduced at the optimal point to ensure timely delivery.

A harder choice lies with whether an item is of standard- or intangible class. My observation is that standard class items typically have opportunity-cost functions that take effect immediately. For example, if we had this new feature today, we could be making money from it tomorrow. So sooner delivery is desirable, but delay does not carry the same penalties as something of a fixed date nature or an expedite request.

Intangible items tend to be associated with important, valuable items that have an (opportunity) cost of delay that does not take effect in the near future. Typically, the point at which the cost function makes an inflection point upward is quarters, or years, in future. It is well beyond the immediate planning horizon, e.g., two or three times the typical lead time. If our current lead time is generally 28 days, then our planning horizon is perhaps three months. Items that will incur a lost opportunity or a tangible cost within that three-month window should be treated as standard class, while items in which the cost or benefit is not realized until quarters or years in the future should be treated as intangible class items.

Putting Classes of Service to Use

Classes of service should be defined for each kanban system. The policies associated with each class of service should be explained to every team member. Everyone attending a standup meeting in the morning should appreciate and understand the classes of service in use. To make this effective, the number of classes of service should be kept fairly small—four to six is a good guideline. And again, because we want each team member to remember the classes of service, what they mean, and how to use them, the number of policies for each class of service should be kept small and simple. The definitions should be precise and unambiguous. Again, a good guideline would be no more than six policies per class of service.

Armed with an understanding of classes of service and knowledge of the policies associated with them, the team should be empowered to self-organize the flow of work. Work items should flow through the system in a fashion that optimizes business value and customer service, resulting in releases of software that maximize customer satisfaction.

Allocate Capacity to Classes of Service

Figure 11.3 shows a kanban system with a total WIP limit of 20. Classes of service are designated by four colors of tickets. The white expedite tickets do not count against the WIP limit, but are limited to only one item at a time. Hence, they have a five percent impact on total capacity when present, and increase the effective work-in-progress to 21 items. In this example, fixed delivery date purple tickets represent 20 percent of the total. This means that there can only be four purple tickets on the board at any given time, but they can be present in any column. Yellow standard class items represent 50

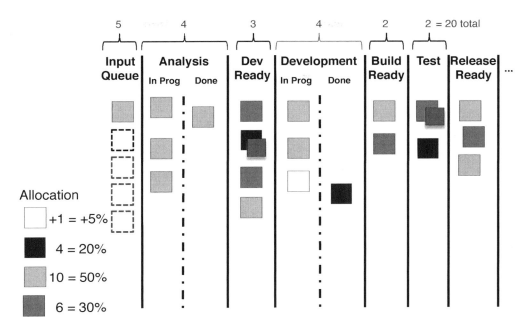

Figure 11.3 Card wall showing capacity allocation across classes of service

percent of the total allocation, for a total of ten tickets. The remaining 30 percent are allocated to green intangible class items.

Now that we have allocated capacity to different classes of service, the input queue replenishment activity is complicated by the available capacity for each class. As currently shown, there is capacity for one fixed delivery date item and three intangible class items. This generates a lot of questions. What if we do not have current demand for a fixed date item? What should we do? Should we fill the slot with a standard class item? If so, should that item be given fixed date status or treated as a standard class item? If we did this, wouldn't it be adjusting the capacity-allocation policy?

All of these are legitimate questions and highlight legitimate issues that come up on a daily basis when using a kanban system. There are no right or wrong answers to these questions. The answers need to be derived in context and are specific to each situation.

What we can deduce from the chosen allocation is that the domain has a significant number of items of a fixed date nature and that there is also a sizeable capacity being set aside for intangible items. This may imply that there are some major initiatives with longer delivery dates—such as a platform replacement—underway. It may also indicate significant risks in the domain. Perhaps there is a seasonal nature to the demand that increases the number of expedite requests or fixed date items. In order to be able to respond gracefully to this seasonal demand without causing increased customer dissatisfaction, we are choosing to allocate more capacity to intangible items over standard class items. We are building more slack into the system.

As for choices about what to do when our input queue has an open fixed date slot when no suitable fixed date class items are available, this depends on the risks present in the domain. If there is significant demand for fixed date items and the costs associated with these items are high (hence risk is high), we may choose to leave the slot empty. It may make sense to effectively reserve capacity for a future fixed delivery date item. However, if risks are low, we may choose to fill the slot with a standard class item. If later a fixed date item comes along, we could choose to bump the standard class item down in class or to exceed the WIP limit temporarily. All of these choices will have different effects on lead time, due-date performance, spread of variability in lead time, customer satisfaction, and risk management. You will need to make these decisions for yourself; it will take some time to develop suitable experience and judgment to enable you to make the best choices for your team, project, or organization.

Capacity allocation is just another strategy of the kanban system. If you find that your allocation is misaligned with demand, then adjust it—change the policies and adjust WIP limits accordingly.

Takeaways

❖ Classes of service offer a shorthand method for optimizing customer satisfaction.

❖ Work items should be assigned to a class of service according to their business impact.

❖ Classes of service should be clearly, visually displayed by using, for example, different colored cards to represent the class of service or different swim lanes on the card wall.

❖ A set of management policies should be defined for each class of service. Only classes of service related to riskier items should involve wasteful activities such as estimation.

❖ Team members should be trained to understand the classes of service and the policies associated with them.

❖ Some classes of service should include a target lead time.

❖ Due Date Performance (percentage) should be monitored for target lead times.

❖ Classes of service enable self-organization, empower team members, and free up management time to focus on the process instead of the work.

❖ Classes of service change the customer psychology.

❖ If classes of service are used properly and combined with a regular delivery cadence, very few complaints are likely to be received, even if a significant portion of items miss their target lead time.

❖ Kanban system capacity should be allocated to each class of service.

❖ Percentage capacity allocated to each class of service should be aligned with demand.

❖ **Chapter 12** ❖

Metrics and Management Reporting

Although the idea is for Kanban to be minimally invasive and change as little as possible of the value stream, job roles, and responsibilities, it does change the way the team interacts with its partners—the external stakeholders. Because of this, Kanban needs to report slightly different metrics than you may be used to with either a traditional or an Agile project-management approach.

Kanban's continuous-flow system means that we are less interested in reporting on whether a project is "on-time" or whether a specific plan is being followed. What's important is to show: that the Kanban system is predictable and is operating as designed, that the organization exhibits business agility, that there is a focus on flow, and that there is clear development of continuous improvement.

For predictability, we want to show how well we perform against the class-of-service promises. Are work items being treated appropriately, and, if the class of service has a target lead time, how well are we performing against that? What is the due-date performance?

For each of our indicators, we want to track the trend over time, so we can see the spread of variation. If we are to demonstrate continuous improvement, we want the mean trend to improve over time. If we are to demonstrate improved predictability, we want the spread of variation to decrease and the due-date performance to increase.

Tracking WIP

Before we get to performance indicators, however, I believe that the most fundamental metric should show that the kanban system is operating properly. To do this, we need a cumulative-flow diagram that shows the quantities of work-in-progress at each stage in the system. If the kanban system is flowing correctly, the bands on the chart should be smooth and their height should be stable.

The example charted in Figure 12.1 shows how well the team is doing at maintaining the WIP limits. We can see that WIP (the middle, light-colored band) is growing in the middle of the time period. At the beginning, the WIP limit is correctly 27. At the end of the period, due to a personnel adjustment, the WIP limit is correctly 21. We can also read the average lead time by scanning this chart horizontally.

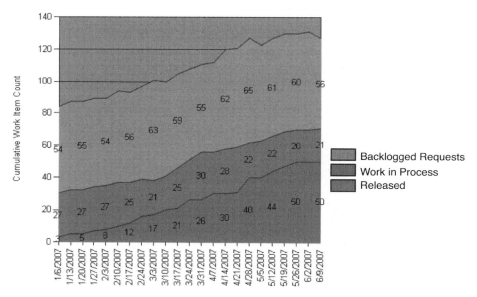

Figure 12.1 Example of Cumulative Flow diagram from a Kanban System

Lead Time

The next metric of interest indicates how predictably our organization delivers against the promises in the class-of-service definitions. The underlying metric for this is lead time. If an item was expedited, how quickly did we get it from the order into production? If it was of standard class, did we deliver it within the target lead time? I've found the best way to show this data is with a spectral analysis of the lead time, which plots the target lead time against the service-level agreement (SLA) for a class of service (see Figure 12.2).

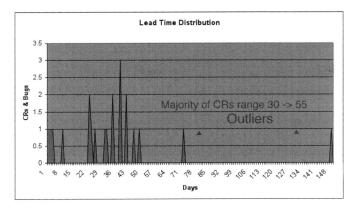

Figure 12.2 Example of Lead Time spectral analysis

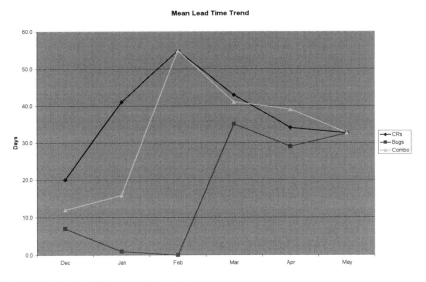

Figure 12.3 Example of Mean Lead Time Trend

Reporting the average (mean) lead time has some use as a report card on overall performance (see Figure 12.3), but it is not very helpful as either an indicator of predictability or a means to inform on improvement opportunities.

The spectral analysis is much more useful because it informs us about items that just barely failed to meet the target time, and of other statistical outliers. In the example shown in Figure 12.4, it would make sense to investigate the root causes of the cluster of items that just failed to meet the target. If these root causes could be addressed, the Due Date Performance (percentage of items delivered as expected) should improve.

Lead Time and Due Date Percentages	Lead Time (Average # of Days)			Due Date Performance (%)	
Interval	Target	May 2007	Dec 2006 to May 2007	May 2007	Dec 2006 to May 2007
Lead Time, Engineering Ready to Release (CRs & Bug Fixes)	30	32.5	31.1	52	50
Lead Time, Engineering Ready to Release (CRs Only)	30	32.6	40.4	50	30
Lead Time, Engineering Ready to Release (Bugs Only)	30	32.5	19.6	55	75

Figure 12.4   Example of report showing mean lead time and due date performance

Due Date Performance

I've found it useful to report Due Date Performance for the most recent month and for the year to date. You may also want to report performance year-on-year (or 12 months ago) for comparison. Hence, it is useful to have 13 months of data.

With the Fixed Delivery Date class of service items, you can include these in the Due Date Performance metric. In this case, you are answering the question, "Was the item delivered on time?" However, although you will have a lead time recorded, that in itself is not as interesting as comparing the estimated lead time to the actual. Estimate versus actual demonstrates how predictable the team is and how well they are performing with Fixed Delivery Date service items. Remember that Fixed Date items receive some analysis and an estimate. The due date performance on fixed date items is a factor in determining the quality of the initial estimate. Naturally, the most important metric is whether the item was delivered on time prior to the hard date. The accuracy of the estimate is an indicator of how efficiently the system is running. If estimates are known to be inaccurate, the team will tend to start fixed date items early in order to assure delivery. This is not optimal. The overall performance in terms of value and throughput can be improved by improving estimation.

Throughput

Throughput should be reported as the number of items—or some indication of their value—that were delivered in a given time period, such as one month. Throughput should be reported as a trend over time, as shown in Figure 12.5. The goal is to continually increase it. Throughput is very similar to the Agile "velocity" metric. It indicates how many user stories, or story points, were completed in a given period. If you are not using Agile requirements techniques but processing other things, such as functional specification items, change requests, use cases, etc., then report the number of those.

Throughput And Production Rate:

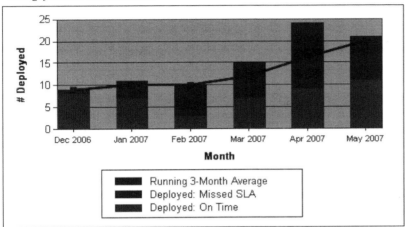

Figure 12.5   Example of bar chart showing throughput

In the first instance it is important to be able to report the raw number. As your team matures and becomes more sophisticated, you may be able to report the relative size, such as the total number of story points, function points, or some other measure of quantity. If your organization is very sophisticated, you may be able to report the value of the work delivered as a dollar amount. At the time of this writing I know of only one team, at the BBC in London, that is capable of reporting the dollar value of work delivered.

Throughput data is used in Kanban for an entirely different purpose than velocity in a typical Agile development environment. Throughput is not used to predict the quantity of delivery in a time interval or any specific delivery commitment. Throughput is used as an indicator of how well the system (the team and organization) is performing and to demonstrate continuous improvement. Commitments in Kanban are made against lead time and target delivery dates. Throughput may be used on larger projects to indicate the approximate time to completion with appropriate buffering for variation.

Issues and Blocked Work Items

The Issues and Blocked Work Items chart shows a cumulative-flow diagram of reported impediments overlaid with a graph of the number of work-in-progress items that have become blocked, as shown in Figure 12.6. This chart gives us an indication of how well the organization is at identifying, reporting, and managing blocking issues and their

How many issues and blocked work items do we have?

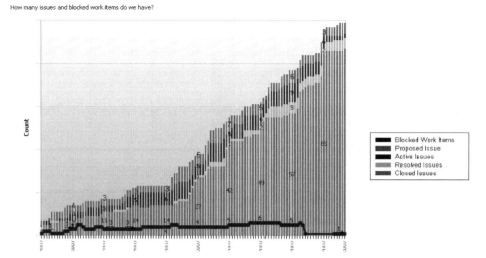

Figure 12.6 Example of Issues and Blocked Work Items chart

impact. If Due Date Performance is poor, there should be corresponding evidence in this chart demonstrating that a lot of impediments were discovered and were not resolved quickly enough. This chart can be used on a day-to-day basis to alert senior management of impediments and their impact. It also can be used as a long-term report card to indicate how capable the organization is at resolving impediments and keeping things flowing—a measure of capability in issue management and resolution.

Flow Efficiency

A good Lean indicator of the waste in the system is to measure the lead time against the touch time. In manufacturing, touch time refers to the time a worker spends actually touching a job. In software development, this is very difficult to measure. However, most tracking systems can track assigned time (to an individual) against time spent blocked and queuing. Hence, although reporting the ratio of lead time to assigned time doesn't give us an accurate indication of the waste in the system, it does give us a conservative ratio that shows how much potential there is for improvement, as in Figure 12.7.

Do not be alarmed if this ratio is initially about 10:1. I've attended many conferences and seen presenters from industries as diverse as new aircraft design and medical equipment design report similar ratios. It seems that with knowledge work we are terribly inefficient and incapable of the agility required to turn an idea or request into a working product efficiently.

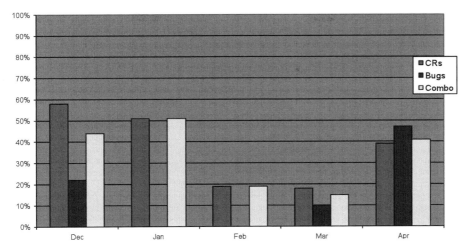

Figure 12.7 Example of lead time: assigned time ratio

The flow-efficiency metric is not very useful day to day, but, again, can be another indicator of continuous improvement.

Initial Quality

Defects represent opportunity cost and affect the lead time and throughput of the Kanban system. It makes sense to report the number of escaped defects as a percentage against the total WIP and throughput. Over time, we want to see the defect rate fall to close to zero, as shown in Figure 12.8.

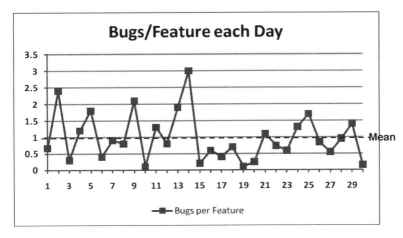

Figure 12.8   Chart showing defects per feature

Failure Load

Failure load tracks how many work items we process because of earlier poor quality—how many work items are production defects or new features that have been requested through our customer-service organization because of poor usability or a failure to anticipate user needs properly. Ideally, failure load should fall over time. This is a good indicator that we are improving as a whole organization and thinking at a system level.

Takeaways

❖ Track WIP with a cumulative-flow diagram to monitor day-to-day WIP limits.

❖ Track lead time for each item processed and report both the mean and the spectral analysis for each class of service.

❖ Lead time is an indicator of business agility.

❖ Track estimated versus actual lead time for Fixed Delivery Date class-of-service items.

❖ Report Due Date Performance as an indicator of predictability.

❖ Impediments block flow and impact lead time and due date performance; report blocking issues and the number of blocked work items in a cumulative-flow diagram with an overlying graph of blocked items. Use it to indicate the capability to report problems and resolve them quickly.

❖ Flow efficiency is the ratio of lead time to assigned engineering time. It indicates how efficient the organization is at processing new work, and it is a secondary indicator of business agility. It also indicates how much room for improvement is available without changing engineering methods.

❖ Initial Quality reports the number of bugs discovered by testers within the system and indicates how much capacity is being wasted through poor initial quality.

❖ Failure Load reports the percentage of work that is generated through some failure of the system and shows capacity left on the table that could be used for new, value-added features.

❖   **Chapter 13**   ❖

Scaling Kanban

So far the examples and stories of Kanban implementations presented here have focused on software maintenance—small system changes made with rapid, frequent releases to production. There are a lot of systems that are in maintenance, and a significant portion of readers involved in software development will find the advice and guidance useful. Equally, there are many more IT personnel involved in support and operations in which ticketing systems for short-order jobs are also common; a Kanban approach would be equally useful for them. However, there are others for whom development of significantly sized projects is the norm. If you are reading this asking why and how would I use Kanban on larger projects and across a project portfolio, I hope that chapter 5 persuaded you that Kanban enables significant and positive cultural changes. The benefits observed with Kanban are sufficiently desirable that they challenge us to ask the question, "How would we do Kanban on large projects?"

Large projects present considerable challenges. Many requirements have to be released together. It will be some considerable number of months before a first delivery is made. The team size is larger. There might be lots of work happening in parallel. Significant pieces of work may need to be integrated. Not all of this work will be software development. For example, documentation and package design may need to be integrated with the final software build before a release can be made.

So how do we deal with these challenges?

The answer is to look to first principles. The first principles of Kanban are to limit work-in-progress and to pull work using a visual signaling system. Beyond that, we look to Lean principles, Agile principles, and the workflow and process that are already in place as our starting position. So we want to limit WIP, use visual controls and signaling, and pull work only when there is capacity to do so; but we also want small batch transfers, to prioritize by value, manage risk, make progress with imperfect information, build a high-trust culture, and respond quickly and gracefully to changes that arrive during the project.

With a large project, just as with a maintenance initiative, you will need to agree on a prioritization cadence for input queue replenishment. The general rule is that a higher cadence with more frequent meetings is better. Look at the principles again. What are the transaction and coordination costs of sitting down with the marketing team or business owners and agreeing on the next items to queue for development? At the other end of the value stream, you will have several integration or synchronization points building toward a release rather than a single release point. So, again from first principles, look at the transaction and coordination costs of any integration or synchronization and agree on a cadence. Again, more often is better. Ask the question, "What is involved in meeting with the business to demo recent work and then integrate it so that it is 'release ready'?"

Next, you will want to agree on WIP limits; the principles for thinking about this do not change. Classes of service will still make sense and help you to cope gracefully with changes during the project.

Hierarchical Requirements

You will also need to define work item types for your project. Many major projects feature hierarchical requirements. It is not uncommon for these to be as many as three levels deep. There may also be different types of requirements, such as customer requirements that came from business owners and product requirements that came from a technical, quality, or architectural team. The requirements might be broken out further into functional and non-functional or quality-of-service requirements. Even with Agile software development, the customer might specify requirements in terms of epic-sized stories that are broken out into user stories and are perhaps broken down further to a lower level of tasks or small units referred to as "grains of sand." I also have seen epics broken into architectural stories that in turn are broken into user stories.

Feature-driven development also has three tiers of requirements—features, feature sets (or activities), and subject areas.

It has made sense for teams adapting Kanban to major projects to set different work item types for different levels of the hierarchy. For example, epic stories are one type of work item, and smaller user stories are another. In a more traditional project, customer requirements are one type, while product requirements are another, and functional-specification items are a smaller, third type.

Typically, teams have chosen to track the two top levels on a Kanban board. I personally have not seen a team or project in which they tried to track three levels with Kanban. Some electronic tools now support hierarchical requirements that allow the user to shell in and out at different levels, displaying only two levels at any given time.

Usually, if there is a third, lower level, such as tasks in an Agile project, the tasks are not tracked on the project card wall or within the team-level kanban system. Individual developers may choose to track tasks, or perhaps small, cross-functional teams might choose to track their tasks locally, but off the larger project board and out of sight of managers and value-stream partners. This isn't motivated by a need to hide information. It is simply that the lowest level of activity isn't interesting from a value-stream and performance-level perspective. The lowest level is often focused on effort and activity rather than on customer value and functionality.

While writing this book, a variant of Kanban for personal use has emerged, championed by Jim Benson and others. Personal Kanban is used at home and in the office at the individual level, or with small groups of two or three who are actively collaborating on the same set of work items. It is impossible to know at the time of this writing whether Personal Kanban will re-assimilate into the wider body of knowledge or emerge as a discipline in its own right.

Decouple Value Delivery from Work Item Variability

What emerged from most Kanban teams tracking the two highest levels of requirements was the idea that the highest level of requirements, the most coarse-grained requirements, were generally describing some atomic unit of value to the market or customer. These epic user stories or customer requirements were often written at a level such that they made sense to release to the market. Were the product already in maintenance, these requests may have been processed and released individually. Sometimes the Kanban community refers to this level of requirements as a "minimal marketable feature" (or MMF). There is some confusion, as MMF was defined by Denne and Cleland-Liang in their book *Software by Numbers,* and their definition isn't strictly adhered to. I would prefer a definition of a minimum marketable release (MMR) that describes a set of features that are coherent as a set to the customer and useful enough to justify the costs of delivery.

It does not make sense to treat an MMR as a single item to flow through our kanban system. An MMR is made up of many work items. MMR makes sense from a release transaction-cost perspective, not from a flow perspective. In some cases a small but highly valuable, differentiated new feature may make economic sense to release. On the other hand, as many have found out, "the first MMF is always large" because the first release of a new system must include all of the essential capabilities to enter a market and all of the infrastructure to support them. There can be two or three orders of magnitude difference in the size of MMFs (or MMRs). A work item type with instances that vary in size up to a thousand-fold will be problematic.

Kanban systems do not appreciate such wide variation in size. They require large buffers and excess WIP to smooth flow; without them they will suffer wide fluctuations in lead time. Large buffers and more WIP mean long lead times and a loss of business agility. The alternative is worse! If we do not buffer for variability in size, there will be wide fluctuations in lead time. As a result, it is impossible to offer a target lead time under a service-level agreement that can be met with any consistency. The result will be poor predictability and a loss of trust in the system. Designing a kanban system around the concept of MMF is likely to lead to a loss of business agility, and/or a loss of predictability, a loss of trust between IT and the business, and general dissatisfaction with Kanban as an approach.

However, using a Minimal Marketable Release (MMR) to trigger a delivery, in conjunction with smaller, fine-grained work item types, is likely to minimize costs and maximize satisfaction with what is released.

Teams can adapt to this challenge by focusing on analysis techniques that produce a lower level of requirements, such as user stories or a functional specification. These will generally be fine-grained, small, and have a relatively small variation in size. An ideal size would be something in the range of a half-day to four days or so of engineering effort.

On one major project, we found that each larger work item, called a "Requirement," and tracked with green tickets, broke out into an average of 21 smaller "Features," tracked with yellow tickets. Although the features were written in a user- and value-centric fashion, they were analyzed to be small and similarly sized. Had this been an Agile project, these two levels might have been Epic, tracked in green, and User Story, tracked in yellow.

The smaller, fine-grained items enable flow and predictability of throughput and lead time, while coarser-grained items at the higher tier in the board allow us to control the number of releasable, marketable requirements in progress at any given time.

By adopting this two-tiered approach, we have decoupled the delivery of value from the variability of the size and effort required to deliver that value.

It makes sense to set WIP limits at both levels. With several projects we found it made sense to assign small, cross-functional teams to each of the higher-level requirements. Those teams would pull all the smaller, finer-grained items for that higher-level requirement and flow them across the board without handoffs until the requirement was complete and ready for integration or delivery. Then the team would pull another coarse-grained requirement. There would also be the opportunity to reassign team members, either adding or releasing individuals from the team, depending on the size of the next item to be pulled.

Two-tiered Card Walls

The first teams using Kanban on large projects adopted a two-tiered style of card wall, as shown in Figure 13.1.

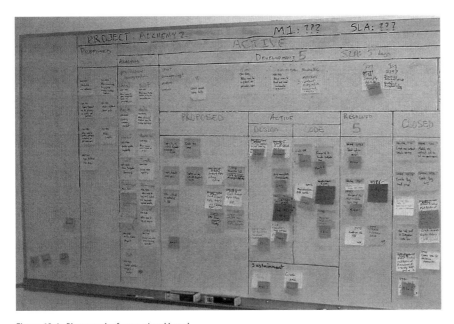

Figure 13.1 Photograph of a two-tiered board

In this picture, the coarse-grained requirements are shown with green tickets. They flow left to right through a set of states, namely, backlog, proposed (analysis), active (design & development), resolved (testing) and closed.

The requirements that are active are shown across the top of the middle section of the picture. They, in turn, are broken out into lots of smaller features, shown with yellow tickets. The features flow through their own set of states, namely, proposed (analysis), active (design and code), resolved (testing), and done. The states the features flow through is similar to the higher-level requirements but they don't need to be. You can choose to model this however you see fit. My advice is to model what you do now. Do not change the process if you can avoid it.

The yellow tickets are linked back to their parents by tagging them with their parents' ID numbers.

In an example like this, it is possible to limit WIP at both levels in the hierarchy, but the yellow tickets are all grouped in one pool. I don't have enough evidence from the field to know whether or not this is a good strategy. What I do know is that it didn't stick with this team.

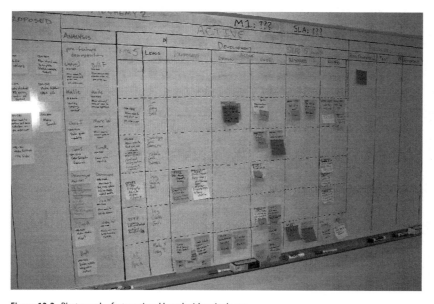

Figure 13.2 Photograph of a two-tiered board with swim lanes

Introducing Swim Lanes

It turns out that matching the finer-grained yellow tickets to their coarser-grained parents is important. It also seems to make sense to limit WIP at the lower level, within an individual cross-functional team. To facilitate this approach, some teams innovated the card-wall system and introduced horizontal swim lanes.

In Figure 13.2, the higher-level requirements, shown with green tickets, flow through the same set of states—namely, backlog, proposed, active, resolved, and closed. However, the middle section has been redrawn compared to Figure 13.1. The active, coarse-grained requirements in green are vertically stacked to the center left. From each of those green tickets extends a swim lane that is divided into the same set of states as for the finer-grained yellow features. The number of swim lanes is now the WIP limit for the coarse-grained customer-marketable requirements, while the WIP limit for the finer-grained features can now be set for each swim lane if the individual teams choose to do so. The column immediately to the right of the vertical stack of green requirements contains the names of the permanently assigned team members. The small orange tickets attached to yellow tickets actually contain the names of specialist floating resources, such as user-experience designers and database architects.

This swim-lane variant of the card wall means that we are now managing customer-marketable WIP vertically, while we are managing WIP on the low-variability features horizontally. This format proved to be very popular and has become typical.

Alternative Approach to Size Variability

Another approach to dealing with variability in size is to create different work item types for different sizes of items. Swim lanes can then be set up for items of each size/type. WIP limits should be set for each column in each lane, i.e., each box on the card wall. Because there is lower variability across each lane (as the items are all similarly sized), each lane should flow relatively smoothly. This is one way of coping with variability without resorting to a two-tiered system.

Incorporating Classes of Service

The two most obvious methods for visually differentiating cards on the wall are by using both color and swim lanes. However, on large projects, each ticket typically has three attributes that we need to communicate: the work item type, the level in the hierarchy, and the class of service. It's worth noting that, in the example (Figure 13.2), the choice to have different work item types for different levels in the hierarchy and then to use both color and swim lanes to designate the level in the hierarchy means that hierarchy is being overloaded with two methods of visualization.

If you do need to communicate class of service in addition to type and level in the requirements hierarchy, it may make sense to use color for class of service. If types are used for purposes other than hierarchy level, for example, to show bugs or defects, or value-add versus failure load, you may want to choose another approach, perhaps introducing an icon or a sticker attached to the card to signify type; or you may prefer to use color for type and an icon or sticker for class of service, e.g., a silver star for an expedite request.

What may just be easier, as we saw evolve at Corbis, is to use color for multiple purposes, such as level in hierarchy, type, and class of service. This approach, in which color does not clearly designate any one attribute, seems to be acceptable to users of the kanban system and is very efficient in terms of available options for visualization.

Systems Integration

On some larger projects, you may have multiple teams working on different components of a system that need to be integrated later. Some of these components may involve hardware or firmware, and may be not amenable to modern continuous-integration techniques. When you have such components that need to be integrated, you need to determine an integration point based on a coarse-grained high-level planning activity. This point should then be treated as a fixed date for delivery of these dependent components. This allows each team to move forward independently with its kanban system, but also to coordinate delivery of dependent items when they are needed. Late delivery of a dependent item causes excessive delay across a whole project. This high cost of delay justifies treating it as a fixed date class-of-service item.

Managing Shared Resources

It is common on larger projects and across project portfolios to share some specialist resources; for example, software architecture, database architecture and administration, user-experience testing, user-experience design, and software security audit. There are three established methods for dealing with these shared resources in Kanban.

In the first method some of the work items have additional, smaller orange tickets attached to them. These smaller tickets show the name of a required shared resource, such as Sandy, the enterprise data architect. At the lowest level of intrusion, this simple act of visualizing the work of the shared resource is often enough to coordinate that individual's workload. If several tickets start appearing with the same name, it can bring up questions about how this person is managing to work on multiple things all at one time. This might be enough to facilitate a policy-change discussion—does all the work need to be seen by that person?—or to escalate it to the next level.

The next level is to recognize that shared resources are not instantly available, and to visualize that by marking items requiring attention from a shared resource (an orange ticket), as blocked until the person is actively working on them. This has the effect of invoking the issue-management and resolution function to resolve the availability of the shared resource. It also has the effect of highlighting to management whether or not this resource's availability is a problem and a potential bottleneck.

The highest level of management for a shared resource is to give that resource its own kanban system. For example, enterprise data architecture could have its own kanban system, so could user experience, and software security, and so on and on. Each team or resource would independently analyze its demand and set work item types based on the source of request, and classes of service based on priority and response required. The demand would be analyzed and policies for capacity allocation set.

At this level, what has emerged is a service-oriented architecture for actually developing the software itself. Each group within the enterprise offers its own set of services that are exhibited as service-level agreements for different classes of service and work item types. Clients of those shared resources then submit work requests to their backlog; those requests are queued and selected for processing, just as described throughout this book. If requests from a given client are not being processed quickly enough, it can open up discussion about whether or not the system is designed correctly, and whether or not policies about capacity allocation and class of service need to change. It may even provide sufficient evidence for realigning or augmenting staffing.

Takeaways

❖ Major projects should follow the core principles of Kanban.

❖ WIP limits, prioritization cadence, delivery cadence, and classes of service are valid techniques for major projects.

❖ Major projects tend to have hierarchical requirements; these levels of hierarchy should be modeled with work item types.

❖ Typically, teams track the top two levels of requirements in the hierarchy on the card wall and limit WIP at one or both levels.

❖ The highest level of requirements typically models customer-marketable requirements that make atomic units that potentially could be released individually.

❖ The second level of requirements is typically written in customer- and user-centric language and is analyzed in such as way as to make the requirements both fine-grained and similarly sized.

❖ This second level of fine-grained requirements facilitates flow by reducing variability in the kanban pull system.

❖ A two-tiered card wall will be required to visualize both levels of requirements that are being tracked.

❖ Swim lanes have become a popular technique for showing hierarchy and to facilitate limiting WIP.

❖ The coarse-grained WIP is limited by the number of swim lanes.

❖ The fine-grained WIP can be limited on each swim lane, if desired.

❖ Typically, small cross-functional teams are assigned to each swim lane.

❖ Shared-resource demand can be visualized with small stickers that are attached to regular work items.

❖ Non-instant availability of shared resources can be highlighted with blocking-issue tickets (pink, in our example) attached to the original work item ticket.

❖ Shared resources should develop their own kanban systems.

❖ A network of kanban systems for shared resources across a portfolio of projects can be thought of as a service-oriented architecture for software development.

❖ Chapter 14 ❖

Operations Review

Ante Meeting

It's 7:30 A.M. on the second Friday in March 2007. I'm at work early because this morning is our department's fourth monthly operations review. I'm joined by Rick Garber, the manager of our software process engineering group. Rick has the job of coordinating the ops review meeting and agenda. He's busy printing out the handout that contains the approximately 70 PowerPoint slides for today's meeting. Once the printing is done, we head over to the Harbor Club in downtown Seattle with a box of 100 handouts. Ops review is scheduled to start at 8:30 A.M. but a hot buffet breakfast is served beginning at 8:00. The invite includes all of my organization and all of my colleague Erik Arnold's organization. However, with some folks in India, some in other parts of the U.S., and always a few who can't make it for personal reasons, we generally get around 80 attendees.

The invite also includes my boss, the CIO of Corbis, and a number of other senior managers, our value-stream partners. The external group that attends in greatest numbers is the Network and Systems Operations team run by my colleague Peter Tutak. They, after all, have to recover failed systems in production, so they feel the pain of our failure most. They also feel the greatest impact when we make new releases to production. So, arguably, they have the most to gain by actively participating.

The group begins to arrive in good time to enjoy their breakfast. The room is on the top floor of a Seattle tower and affords us all beautiful views of the city, the harbor, the piers, and Elliott Bay. The room is laid out with round tables, with six to eight people seated at each. We have a projector screen and a lectern at one end. Rick manages the schedule with precision. Each presenter has around eight minutes for their four or five slides. There are a few time buffers to allow for the variability that comes with questions and discussion. I kick things off promptly with a few opening remarks. I ask everyone to think back to the end of January and what we were doing back then. I remind everyone that we are here to review the organization's performance for the month of February. Rick has picked out a nice picture from the company archives to symbolize a theme for the month and to help jolt memories, reminding everyone of a key activity from the month.

Set a Business Tone from the Beginning

I hand off the proceedings to Rick, who summarizes the management action items from last month and gives an update on the status. Next we introduce our finance analyst, who presents a summary of the company performance for the month—the reason for delaying until the second Friday of the subsequent month is so that we can have the financial data after the books for the prior month are closed. She summarizes the budget details for both my and Erik's cost centers. We look at planned versus actual for all major budget areas, as well as headcount targets. We discuss open requisitions and encourage team members to submit candidates for open positions. Coming out of this first segment, everyone attending knows how well the company is doing and how well the software engineering group is managing against budget, and therefore, how much slack we have to buy items like flat-screen monitors and new computers. The purpose of leading with the financial numbers is to remind everyone on the team that we are running a business; we are not just showing up each day to have fun with ones and zeroes with a group of friends.

Inviting Guests Broadens the Audience and Adds Value

The next speaker is a guest—a vice president from another part of the company. I had the bright idea that if we wanted our value-stream partners to take an interest, we should show an interest in them and invite them to present. We offered our guest 15 minutes, and he took it. So we heard a presentation on sales operations, the part of the business that fulfills customer orders and ensures delivery of product. Although some of Corbis's business is done on the web and fulfilled electronically, not everything the firm offers is delivered as a download; a whole department fulfills more complex orders for professional advertising agencies and media firms. My colleague, Erik Arnold, had the bright idea of asking the guest to sponsor our breakfast to keep our costs under control. That worked, too. Over the next few months our team learned about many aspects of the business and senior leaders throughout the company learned what we did, how we did it, and how hard we were trying to deal with our issues. Nine months later, executives were openly talking about how well governed the IT team was and how their business unit ought to be following our lead.

Main Agenda

Once our guest speaker finished, we moved on to the main portion of the meeting. Each manager had eight minutes for a presentation on their department's performance. We followed this with some project-specific updates from our program-management office. Each of the immediate team managers got up and spent five minutes quickly presenting their metrics. Generally, they followed the format laid out in chapter 12: They presented information on defect rates, lead time, throughput, value-added efficiency, and, occasionally, a specific report that would drill into some aspect of our process on which they needed more information. Then they took questions, comments, and suggestions from the floor for a few minutes.

This fourth month of the Ops Review, March 2007, was particularly interesting. The first Ops Review was in December. Everyone came, almost 100 percent turnout. Lots of curiosity, and afterward lots of comments like, "I have never seen transparency like this in my career," and "That was very interesting." The most useful piece of feedback was, "Next time can we have a hot buffet rather than cold?" So we added hot breakfast. The second month people said, "Yes, another good month. Somewhat interesting! Thanks for the hot breakfast!" On the third month some of the developers were asking, "Why do I need to get up so early?" and "Is this a good use of my time?"

What happened in the fourth month is that we reviewed a significant problem. The company had acquired a business in Australia. IT had been asked to switch off all the Australian subsidiary's IT systems and migrate all 50 users to the Corbis systems. The request had an arbitrary but urgent date. This date was based on an "economy of scale"-style cost savings that had partly justified the acquisition price, so there was a cost of delay involved. The request had arrived as a single item in our maintenance queue. It was big enough to have justified ten tickets, but we treated it as only one. The effect of an outsized item like this entering a kanban system is well understood in industrial engineering. It clogs the system and greatly extends the lead time for everything that comes in behind it. And so it was with us. Lead time jumped, on average, from 30 to 55 days. Queuing theory also tells you that reducing a backlog when fully loaded takes a long time. We discovered that it would take five months to recover the lead-time target.

In addition, we had a release that had required an emergency fix.

All of a sudden the room was alight with questions, comments, and debate. After three months of boring, good data, we had a story to tell. The staff were amazed that we (the management) were willing to talk openly about the problems and what to do about them, and that Ops Review wasn't only about showing off how good we were, just presenting the good data. None of the staff questioned again why we held the meeting every month.

The meeting ended with Rick summarizing the management action items from the morning's discussions and thanking everyone for coming. It was 10:30 and time to head back across the street to the office.

Keystone of Lean Transition

There are a lot of important things to understand about Operations Review. First, I believe that Ops Review is the lynchpin, or keystone, of a Lean transition and Kanban method implementation. It is an objective, data-driven retrospective on the organization's performance. It is above and beyond any one project and it sets an expectation of objective, data-driven, quantitative management rather than the more subjective, anecdotal, qualitative management that is the more established practice with Agile project and iteration retrospectives. Operations review provides the feedback loop that enables growth of organizational maturity and organization-level continuous improvement. I truly believe that is essential to delivering a successful enterprise-scale Lean (or Agile) transition.

Appropriate Cadence

I also believe that Operations Reviews have to be monthly. More often can be burdensome for data collection, and the time involved for the meeting means that there is a desire not to do it too often. Fitting such a meeting into two hours is challenging. If it were not a data-driven meeting, full of charts and reports, it would not be possible. A subjective, anecdotal-style meeting at that scale could not be completed within two hours. A typical project retrospective takes longer than two hours, so imagine trying to perform an organization-wide retrospective and complete it in two hours using a pluses- and deltas-style analysis. Part of the secret to keeping the length of the meeting short is to hold it based on objective data; keep the agenda tight and manage it throughout the meeting.

There can be a tendency to want to hold operations reviews less often; quarterly is common. My experience with quarterly operations reviews dates from my time with Motorola's PCS division. My observation of those meetings is that they were upper-management reporting-and-review sessions, not organizational sessions designed to drive continuous improvement and organizational maturity. Quarterly is too seldom to really drive an improvement program. The data is often four months old by the time it hits a quarterly review. A quarter is a long time span to review in a single meeting, so the review tends to be superficial. Reports and metrics tend to be of lagging indicators and focused around reporting performance against target to senior leaders.

Quarterly meetings seem attractive because they feel more efficient—just one two-hour meeting every quarter rather than every month. They also cost less on an annual basis—just four meetings rather than 12. After I left Corbis at the beginning of 2008, my former boss reduced the cadence of the operations reviews to quarterly to save money. After three quarters, and with that boss also gone, the new leadership questioned the value of the meetings and decided to cancel them altogether. Within another few months the performance of the organization had allegedly depreciated considerably, and the level of organizational maturity had reportedly fallen back from approximately the equivalent of CMMI Model Level 4 to CMMI Model Level 2, from Quantitatively Managed to merely Managed.

We can draw several things from this. The loss of a feedback loop reduced the opportunities for reflection and adaptation that could lead to improvements. Eliminating a meeting focused on objective performance review of the organization sent a message that leadership no longer cared about performance. The result was a significant step backward in organizational maturity and performance in terms of predictability, quality, lead times, and throughput.

Demonstrating the Value of Managers

The operations review also shows the staff what managers do and how management can add value in their lives. It also helps to train the workforce to think like managers, and to understand when to make interventions and when to stand back and leave the team to self-organize and resolve its own issues. Operations review helps to develop respect between the individual knowledge workers and their managers and between different layers of management. Growing respect builds trust, encourages collaboration, and develops the social capital of the organization.

Organizational Focus Fosters Kaizen

While individual project retrospectives are always useful, an organization-wide ops review fosters institutionalization of changes, improvements, and processes. It encourages improvements to spread virally across an organization and creates a little intramural rivalry between projects and teams that encourages everyone to improve their performance. Teams want to demonstrate how they can help the organization with better predictability, more throughput, shorter lead times, lower costs, and higher quality.

An Earlier Example

I didn't invent operations reviews. They are quite common at many large companies. However, I learned how to do them in this objective, business-unit-wide fashion while I was working at Sprint PCS in 2001. My boss, the Vice President and General Manager of sprintpcs.com, instituted them for very similar reasons. He wanted to develop the maturity of his organization—a 350-person business unit responsible for the web site, all e-commerce, and online customer care for Sprint's cellular telephone business. At sprintpcs.com we held the ops review every third Friday of the month at 2 P.M. It lasted for two hours and brought together around 70 senior staff and managers from the business unit, plus director- or senior manager-level invitees from our upstream and downstream partners. Senior leaders, including the Chief Marketing Officer and the VP of Strategic Planning were also regular attendees. The format was very similar to the one at Corbis. It was entirely objective-data driven. Each manager presented his or her own data. The meeting led off with financial data first. The schedule was planned and managed tightly. After the meeting, everyone got to go home early on a Friday. The meeting was held offsite, at a local college campus. While sprintpcs.com had struggled with Agile software development techniques, the operations review was a key element in developing the organizational maturity and improving the governance of the organization. It showed staff that managers were making a difference and knew how to manage, and it gave staff and line managers a chance to show senior leaders how they could help and where they needed interventions to truly make a difference.

Given two experiments over a period of four years throughout the last decade, I've become convinced that the operations review is a critical piece of a successful Lean or Agile transition and a vital component in developing organizational maturity.

Takeaways

❖ Operations reviews should be organization wide.

❖ Operations reviews should focus on objective data.

❖ Each department should report its own data.

❖ Presentations should be kept short and typically should report metrics and indicators similar to those discussed in chapter 12.

❖ Leading with financial information underscores that the software engineering function is part of a wider business, and that good governance is important.

❖ A monthly cadence for operations reviews seems to be about right. More often is burdensome in both time commitment and in data gathering and preparation. Less often tends to reduce the value and undermine the nature of the meeting.

❖ Meetings should be kept short, typically two hours.

❖ Operations reviews should be used to provide a feedback loop and drive continuous improvement at the enterprise or business-unit level.

❖ Operations reviews show individual contributors how management can add value in their lives and what effective managers do.

❖ Effective operations reviews build mutual trust between the managers and the workers.

❖ External stakeholders attending operations reviews get an opportunity to see how the software engineering and IT groups function and to understand their issues and challenges. This fosters trust and collaboration.

❖ Operations reviews should examine bad data and problems just as much as they should bask in success and extol the virtues of teams with good results.

❖ Holding meetings offsite seems to help focus the attendees' minds.

❖ Providing food appears to encourage attendance.

❖ The involvement of senior leaders communicates that the organization takes performance and continuous improvement seriously.

❖ Signaling a serious interest in performance, continuous improvement, and quantitative management is vital in developing a kaizen culture among the general workforce.

❖ Operations reviews have been shown to lead directly to increased levels of organizational maturity.

❖ Improvement suggestions should be captured as management action items and reviewed for progress at the beginning of the next and subsequent meetings.

❖ Managers should be held accountable and should demonstrate follow through on suggestions.

❖   **Chapter 15**   ❖

Starting a Kanban Change Initiative

Getting started with Kanban isn't typical of process initiatives you may have undertaken in the past. It's important to lay the foundations for long-term success. To do that, it's necessary to understand the goals behind using the Kanban approach to change. I subtitled this book, "Successful Evolutionary Change for Your Technology Business." I did this to underscore the point that the main reason for adopting Kanban is change management. Everything else is secondary.

Cultural Change Rather Than a Managed-Change Initiative

Chapter 5 describes how Kanban optimizes the existing process through a series of incremental, evolutionary changes. This process of optimizing what already exists leads to an improvement in organizational maturity and eventually enables larger, more strategic changes to be introduced. Because of this, it is unlikely that you will drive adoption of Kanban through a planned-transition initiative and a prescribed training program. This is a significant shift away from how a typical Agile transition is planned and managed. In fact, the change-management approach for the introduction of Agile methods is fairly typical of change-management initiatives that have gone before, such as those based on CMMI or the introduction of methods such as the Rational Unified Process. The change initiative tends to be a big-planning-up-front-style initiative. There is a specific type of managed change in which the current process is first defined and assessed, followed by the selection of an Agile method from a textbook. Next, a training and coaching engagement is planned to transition the team or organization from what they do now to the newly defined Agile process. Once this is completed and the new process is in place, another assessment is conducted to demonstrate the adoption of the new methods. This is not the approach with Kanban. With Kanban, there is no planned initiative, no assessments, and no declaration at the end that, "Now we are Agile!" Ideally, there is no end. Instead, leadership drives a continuous process, encouraging incremental changes. As a result, there is a gradual transformation toward a Kaizen culture.

It is true that some training will be necessary. Team members and other stakeholders must understand basics such as the relationship between WIP and lead time and that strictly limiting the quantity of WIP will improve the predictability of lead time. It may also be necessary to provide a brief overview of likely improvement opportunities such as bottlenecks, waste, and variability. As these opportunities for improvement are uncovered, more training in new skills and techniques may be needed. For example, if defects are a major source of waste, the development team may require training in techniques that will greatly reduce defects and improve code quality, such as continuous integration, unit testing, and pair programming.

However, rather than waste too much time on education, in the first instance, it is more important that you gain a consensus around the introduction of Kanban and start using it. This chapter seeks to lay out the foundations for a successful Kanban transition and provides you with a simple 12-step guide to getting started.

Although our main goal with Kanban is to introduce change with minimal resistance, there must be other goals. Change for the sake of change is pointless. These other goals should reflect genuine business needs such as high-quality, predictable delivery. The goals listed here are intended as examples. The specific goals for your organization may differ. Step 1 in your process should be to agree on the goals for introducing Kanban into your organization.

The Primary Goal for Our Kanban System

We are doing Kanban because we believe it provides a better way to introduce change. Kanban seeks initially to change as little as possible. So change with minimal resistance must be our first goal.

Goal 1   Optimize Existing Processes

Existing processes will be optimized through introduction of visualization and limiting work-in-progress to catalyze changes. As existing roles and responsibilities do not change, resistance from employees should be minimal.

Secondary Goals for Our Kanban System

We know that Kanban allows us to deliver on all six elements of the Recipe for Success (from chapter 3). However, we might want to reword the goals slightly and expand some of the points in the recipe to reflect that one point can help us deliver on more than one goal.

Goal 2   Deliver with High Quality

Kanban helps us focus on quality by limiting work-in-progress and allowing us to define policies around what is acceptable before a work item can be pulled to the next step in the process. These policies can include quality criteria. If, for example, we set a strict policy that user stories cannot be pulled into acceptance test until all other tests are passing and bugs resolved, we are effectively "stopping the line" until the story is in good enough condition to continue. With a team new to Kanban, we may not implement such a strict rule, but there should be some policies relating to quality that focus the team on developing working code with low defect numbers.

Goal 3   Improve Lead Time Predictability

We know that the amount of WIP is directly related to lead time and that there also is a correlation between lead time and a non-linear growth in defect rates.* So it makes sense that we want to keep WIP small. It makes our

---

*   At the time of this writing academics are beginning to investigate this relationship between lead time and defect-insertion rates. My hope is that some academic papers will be published in 2010 that will validate my belief that lead time affects defect rates in a non-linear fashion.

lives easier if we simply agree to limit it to a fixed quantity. This should make lead times somewhat dependable and help us to keep defect rates low.

Goal 4   Improve Employee Satisfaction

Although employee satisfaction often gets lip service in most companies, it is seldom a priority. Investors and senior managers all too often take the view that resources are fungible and easily replaced. This reflects a cost-centric bias in their management or investment approach. It doesn't take into account the huge impact on performance that comes with a well motivated and experienced workforce. Staff retention is important. As the population of software developers ages, they care more about the rest of their lives. Many lament wasting their twenties locked up in an office slaving over a piece of code that failed to reach market expectations and became obsolete soon after release.

Work/life balance isn't only about balancing the number of hours someone spends at work with the number of hours they have available for their family, friends, hobbies, passions, and pursuits. It is also about providing reliability. Say, for example, that a team member with a passion for art wants to take a painting class at the local middle school. It starts at 6:30 p.m. and runs every Wednesday for ten weeks. Can your team provide certainty to that individual that he or she will be free to leave the office on time each week in order to attend the class?

Providing a good work/life balance will make your company a more attractive employer in your local market. It will help to motivate employees and it will give your team members the energy to maintain high levels of performance for months or years. It's a fallacy that you get top performance from knowledge workers when you overload them with work. It might be true tactically for a few days, but it isn't sustainable beyond a week or two. It's good business to provide a good work/life balance by never overloading your teams.

Goal 5   Provide Slack to Enable Improvement

Although the third element of the Recipe for Success—Balance Demand against Throughput—can be used to avoid overloading team members and to allow them a reliable work/life balance, it has a second effect. It creates slack in the value chain. There must be a bottleneck in your organization.

Every value chain has one. The throughput delivered downstream is limited to the throughput of the bottleneck, regardless of how far upstream it might be. Hence, when you balance the input demand against the throughput, you create idle time at every point in your value chain except at the bottleneck resource.

Most managers balk at the idea of idle time. They've generally been trained to manage for utilization (or efficiency, as it is often called), and inherently it feels like changes can be made to reduce costs if there is idle time. This may be true, but it is important to appreciate the value of slack.

Slack can be used to improve responsiveness to urgent requests and to provide bandwidth to enable process improvement. Without slack, team members cannot take time to reflect upon how they do their work and how they might do it better. Without slack they cannot take time to learn new techniques, or to improve their tooling, or their skills and capabilities. Without slack there is no liquidity in the system to respond to urgent requests or late changes. Without slack there is no tactical agility in the business.

Goal 6   Simplify Prioritization

Once a team is capable of focusing on quality, limiting WIP, delivering often, and balancing demand against throughput, they will have a reliable, trustworthy, software development capability: an engine for making software! A "software factory" if you will. Once this capability is in place, it would behoove the business to make optimal use out of it. To do this requires a prioritization method that maximizes business value and minimizes risk and cost. Ideally, a prioritization scheme that optimizes the performance of the business (or technology department) is most desirable.

The software-engineering and project-management fields have been developing prioritization schemes since software projects began, perhaps 50 years ago. Most of the schemes are simple. For example, High, Medium, and Low provide three simple classifications. None of these has any direct meaning for the business. Some more elaborate schemes came into use with the arrival of Agile software development methods such as MoSCoW ("Must have," "Should have," "Could have," "Won't have"). Other methods such as Feature Driven Development used a modified and simplified version of the Kano Analysis technique popular among Japanese companies. Yet others advocated strict enumerated order (1, 2, 3, 4…) by business value or technical risk. The challenge with this latter scheme is that it often creates a conflict between the high-risk items that should be prioritized first and the high-value items that also should be prioritized first.

All of these schemes suffer from one fundamental problem. In order to respond to change in the market and evolving events, it is necessary to reprioritize. Imagine, for example, that you have a backlog of 400 requirements prioritized in a strict enumerated order—1 to 400—and you are delivering incrementally using an Agile development method in one-month iterations. Every month you will have to reprioritize the remaining backlog of up to 400 items.

In my experience, asking business owners to prioritize things is challenging. The reason for this is simple: There is so much uncertainty in the marketplace and the business environment. It's hard to predict the future value of one thing against another, when something might be needed, and whether something else might be more valuable to have sooner. Asking a business owner to prioritize a backlog of technology system requirements is to ask them a very hard set of questions to which the answers are uncertain. When people are uncertain, they tend to react badly. They may move slowly. They may refuse to cooperate. They may become uncomfortable and dysfunctional. They may simply react by thrashing and constantly changing their minds, randomizing project plans, and wasting a lot of team time reacting to the changes.

What is needed is a prioritization scheme that delays commitments as long as possible and that asks a simple question that is easy to answer. Kanban provides this by asking the business owners to refill empty slots in the queue while providing them with a reliable lead-time and due-date performance metric.

We already have six lofty and valuable goals for our Kanban system, and for many businesses, that might be enough. However, I and other early adopters of Kanban have discovered that two other even loftier goals are both possible and desirable.

Goal 7 Provide a Transparency on the System Design and Operation

When I first started to use Kanban systems, I believed in transparency onto work-in-progress, the delivery rate (throughput), and the quality because I understood that it built trust with customers and more senior management. I was providing transparency onto where a request was within the system, when it might be finished, and what quality was associated with it. I was also providing transparency into the performance of the team. I did this to provide customers with confidence that we were working on their request and with an idea of when it might be completed. In addition, I wanted to educate senior management on our techniques and performance and build their confidence in me as a manager and my team as a well-formed professional group of software engineers.

There is a second-order effect from all of this transparency that I hadn't predicted. While transparency onto work requests and performance is all very well, transparency into the process and how it works has a magical effect. It lets everyone involved see the effects of their actions or inactions. As a result, people are more reasonable. They will change their behavior to improve the performance of the whole system. They will collaborate on required changes in policy, personnel, staff resourcing levels, and so forth.

Goal 8   Design a Process to Enable Emergence of a "High-Maturity" Organization

For most senior business leaders that I speak to, this final goal really represents their wishes and expectations for their businesses and their technology development organizations. They seek predictability above all else, coupled with business agility and good governance.

Business leaders want to be able to make promises to their colleagues around the executive committee table, to their boards of directors, to their shareholders, to their customers, and to the market in general, and they want to be able to keep those promises. Success at the senior-executive level depends a lot on trust, and trust requires reliability. Above all else, senior leaders want risk managed appropriately so that they can deliver predictable results.

In addition, they recognize that the world today is fast paced and that change happens rapidly: New technologies arrive; globalization changes both labor markets and consumer markets, causing huge fluctuations in demand (for product) and supply (of labor); economic conditions change; competitors change their strategies and market offerings; and market tastes change as the population ages and becomes wealthier and more middle class. So business leaders want their businesses to be agile. They want to respond to change quickly and take advantage of opportunities.

Underlying all of this they want good governance. They want to show that investors' funds are being spent wisely. They want costs under control and they want their investment-portfolio risk spread optimally.

To do all of this, they'd like to have more transparency into their technology-development organizations. They'd like to know the true status of projects and they'd like to be able to help when it is appropriate. They want a more objectively managed organization that reports facts with data, metrics, and indicators, not anecdotes and subjective assessment.

All of these desires equate to an organization operating at what the SEI defines as Maturity Level 4 on its 5-point scale of capability and maturity in the CMMI. Level 4 and 5 on this scale are known as the high-maturity levels. Very few organizations have achieved this level of maturity regardless of whether or not they have sought a formal Standard CMMI Appraisal Method for Process Improvement (SCAMPI) appraisal. It is no wonder then, that most senior leaders of large technology companies are frustrated by the performance of their software engineering teams because the actual organizational maturity does not match the desired level.

Know the Goals and Articulate the Benefits

So now we have a set of goals for our Kanban system. We need to know these goals and to be able to articulate them, because before we start with Kanban, we need to gain agreement with the stakeholders in our value chain. Kanban will change the way we interact with other groups in the business. If these stakeholders are to accept the changes, we must be able to articulate the benefits.

What follows is a prescriptive step-by-step guide to bootstrapping a Kanban system for a single value chain in your organization. This guide has been developed based on real experience and validated by several early adopters of Kanban, both those who followed (roughly) these steps and were successful and those who recognized that their partial failure could have been prevented had this guide been available at the time.

This guide is provided, in part, to draw attention to the difference between Kanban and earlier Agile development methods. Kanban requires a collaborative engagement with the wider value chain and middle (and perhaps senior) management right from the start. A unilateral, grassroots adoption of Kanban without first building a consensus of managers external to the immediate team will have only limited success and deliver limited benefits to the business.

It has been pointed out to me that this set of steps can seem daunting, and some people have remarked that had they read this, they might have been put off trying Kanban altogether. I hope that the wider scope of this book has explained how to engage with each of these steps and has provided you with useful advice learned from experience in the field.

Steps to Get Started

1. Agree on a set of goals for introducing Kanban.

2. Map the value stream (the sequence of all the actions the development organization carries out to fulfill a customer's/stakeholder's request). (See chapter 6.)

3. Define some point where you want to control input. Define what is upstream of that point and who the upstream stakeholders are (explained in chapter 6). For example, do you wish to control requirements arriving to the design team pre-production? The upstream stakeholders might be product managers.

4. Define some exit point beyond which you do not intend to control. Define what is downstream of that and who the downstream stakeholders are (explained in chapter 6). For example, maybe you don't need to control the delivery fulfillment of the product.

5. Define a set of work item types based on the types of work requests that come from the upstream stakeholders (explained in chapter 6). Do you have some item types that are time-sensitive and others that are not? If so, then you may require some classes of service (explained in chapter 11).

6. Analyze the demand for each work item type. Observe the arrival rate and variation. Is the variation seasonal or event-driven? What risks are associated with this type of demand? Should the system be designed to cope with average or peak demand? What is the tolerance for late or unreliable delivery of this type of work? Create a risk profile for the demand. (Explained in chapter 6.)

7. Meet with the upstream and downstream stakeholders—this might be one big meeting, or it might be lots of little meetings. (Explained in more depth later in this chapter.)

    a. Discuss policies around capacity of the piece of the value stream you want to control and get agreement on a WIP limit (explained in chapter 10).

    b. Discuss and agree on an input-coordination mechanism, such as a regular prioritization meeting, with the upstream partners (explained in chapter 9).

    c. Discuss and agree on a release/delivery-coordination mechanism, such as a regular software release, with the downstream partners (explained in chapter 8).

    d. You may need to introduce the concept of different classes of service for work requests (explained in chapter 11).

    e. Agree on a lead-time target for each class of service of work items. This is known as a service-level agreement (SLA), and is explained in chapter 11.

8. Create a board/card wall to track the value stream you are controlling (explained in chapters 6 and 7).

9. Optionally, create an electronic system to track and report the same (explained in chapters 6 and 7).

10. Agree with the team to have a standup meeting every day in front of the board; invite up- and downstream stakeholders but don't mandate their involvement (explained in chapter 7).

11. Agree to have a regular operations review meeting for retrospective analysis of the process; invite up- and downstream stakeholders but don't mandate their involvement (explained in chapter 14).

12. Educate the team on the new board, WIP limits, and the pull system. Nothing else in their world should have changed. Job descriptions are the same. Activities are the same. Handoffs are the same. Artifacts are the same. Their process hasn't changed, other than your asking them to accept a WIP limit and to pull work based on class-of-service policy rather than receiving it in a push fashion.

Kanban Strikes a Different Type of Bargain

Kanban requires the software development team to strike a different bargain with its business partners. To understand this, we must first understand the typical alternatives that are in common use.

Traditional project management makes a promise based on the triple constraint of scope, schedule, and budget. After some element of estimation and planning, a budget is set aside to provide resources, and a scope of requirements and schedule are agreed upon.

Agile project management, meanwhile, doesn't make such a bold, rigid commitment. There may be an agreed-upon delivery date some months in the future, but the precise scope is never stated. Some high-level definition of scope may be in place, but fine details are never locked down. A budget (or burn rate) may have been agreed upon in order to provide a fixed amount of resources. The Agile development team proceeds in an iterative fashion, delivering increments of functionality in short, time-boxed iterations (or sprints.) Typically, these are one to four weeks in length. At the beginning of each of these iterations, some planning and estimation are performed and a commitment is made. The scope is prioritized often and it is understood that if the team cannot make the commitment, scope is what will be dropped, and the delivery date will be held constant. At the iteration (or time-box) level, Agile development looks very similar to traditional project management. The only key difference is the explicit understanding that scope will be dropped if something has to give; whereas a traditional project manager may choose to slip the schedule, add resources, drop scope, or some combination of all three.

Kanban strikes a different type of bargain. Kanban does not seek to make a promise and commit based on something that is uncertain. A typical Kanban implementation involves agreement that there will be a regular delivery of high-quality working software—perhaps every two weeks. The external stakeholders are offered complete transparency into the workings of the process and, if they want, daily visibility of progress. Equally, they are offered frequent opportunities to select the most important new items for development. The frequency of this selection process is likely to be more frequent than the delivery rate—typically, once per week, though some teams have achieved on-demand selection or very frequent rates, such as daily or twice a week.

The team offers to do its best work and deliver the largest quantity of working software possible; and to make ongoing efforts to increase the quantity, frequency, and lead time to delivery. In addition to offering the business incredible flexibility to select items for processing in very small quantities, the team may also offer the business additional flexibility on priority and importance by offering several classes of service for work. This concept is explained in chapter 11.

Kanban does not offer a commitment on a certain amount of work delivered on a certain day. It offers a commitment against the service-level agreements for each class of service, underpinned with a commitment to reliable regular delivery, transparency, flexibility on prioritization and processing, and continuous improvement on quality, throughput, frequency of delivery, and lead time. Kanban offers to a commitment to a level of service, balancing risk through aggregation across a larger quantity of items. A kanban system properly designed offers a commitment to things that customers truly value. In exchange, the team asks for a long-term commitment from the customers and value-chain partners: a commitment to have an ongoing business relationship in which the software development team strives constantly to improve the level of service through improved quality, throughput, frequency of delivery, and lead time to delivery. Because the customer recognizes an ongoing, long-term relationship, if the customer is willing to measure the service level rather than hold the team to precision on any one item, the system can be made to work.

The traditional approach to forming a commitment around scope, schedule, and budget is indicative of a one-off transaction. It implies that there is no ongoing relationship; it implies a low level of trust.

The Kanban approach is based on the notion that the team will stay together and engage in a supplier relationship over a long period of time. The Kanban approach implies lots of repeat business. It implies a commitment to a relationship, not merely to a piece of work. Kanban implies that a high level of trust is desired between the software team and its value-stream partners. It implies that everyone believes that they are forming a long-term partnership and that they want that partnership to be highly effective.

A Kanban commitment is asking everyone in the value chain to care about the performance of the system—to care about the quality and quantity of software being delivered, the frequency of delivery, and the lead time to deliver it. Kanban asks the value-chain partners to commit to the concept of true business agility and to agree to work collaboratively to make it happen. This significantly differentiates Kanban from earlier Agile approaches to software development.

By taking the time to establish the Kanban bargain with up- and downstream stakeholders, you are establishing an underlying commitment to system-level performance. You are establishing the foundation for a culture of continuous improvement.

Striking the Kanban Bargain

A critical piece of a successful Kanban implementation is the initial negotiation of this different type of bargain. What's going on during these initial negotiations is the establishment of the rules of the collaborative game of software development that will be played going forward. It's vital that value-stream partners are involved with setting up these rules, because it will be necessary for them to stick to them if the game is to be played fairly and the outcome reflect the goals and intent.

Step 7 in our 12-step process for introducing Kanban suggests that we meet with upstream stakeholders, such as marketing or business people who provide requirements, and downstream partners, such as systems operations and deployment teams or sales and delivery organizations. We need to agree with them on policies around WIP, prioritization, delivery, classes of service, and lead time. The set of policies we agree on with these partners will define the rules of our collaborative game of software development. It's hard to treat each of the five elements in isolation as they are essentially interrelated. So although we understand that we must set policy around each of the five elements, the negotiations are likely to be quite circular in nature as the participants iterate on options. For example, if a proposed lead-time target is unacceptable, it may be possible to introduce a different class of service that offers a shorter lead time for certain types of work requests. The five elements—WIP, prioritization, delivery, classes of service, and lead time—provide levers that can be pulled to affect the performance of the system. The skill is in knowing how to pull those levers and how to trade off options to devise an agreement that will work effectively.

WIP Limits

I met a development manager in Denmark who told me that his developers work on seven and one-half tasks, on average, simultaneously. This is clearly undesirable. I wonder if anyone would truly believe that this level of multi-tasking is appropriate. If I were him, I would use this fact as a starting point for my negotiations. I'd open up the conversation by stating that on average, team members were working on seven and one-half things in parallel. I would point out what this does to lead time and predictability and I'd invite my colleagues—the other stakeholders—to suggest what a better number might be. Some of them might suggest that only one item per person is the best idea. And it may be, but it's a very aggressive choice. What if something got blocked? Would it not be good to have an alternative to switch to? So perhaps another person might suggest that two things in parallel is the right answer. Some might argue for three; the range of suggestions is likely to lie between one and three. If the team has ten developers and you can gain a consensus around a maximum of two things in process per person then you have an agreement on a WIP limit of 20 for the development team.

There are other alternatives. Perhaps you want teams to work in pairs of programmers; so two things per pair with ten developers would mean a WIP limit of ten. Alternatively, you may be using a highly collaborative method such as Feature Driven Development or Feature Crews, in which small teams of up to five or six people work on single Minimum Marketable Features, User stories, or batches of Features (as in FDD), known as a Chief Programmer Work Package (CPWP). An FDD team may agree to limit CPWPs to three across a team of 10 developers. (A CPWP is typically optimized for development efficiency based on architectural analysis of the domain and contains from 5 to 15 very fine-grained functions.)

So we've had a conversation about the WIP limit with our stakeholders. We did this by discussing what a reasonable expectation might be for multi-tasking and relating it to reliability of delivery and lead-time expectations. Getting our partners to agree on the WIP limits is a vital element. Although we could unilaterally declare WIP limits, by involving other stakeholders and forming a consensus, we establish a commitment to the rules of our collaborative game. At some point in the future, this commitment will prove invaluable. There will be a day when our partners ask us to take on some additional work. They will do this because something is important and valuable. Their reasons and motives will be genuine. When they do so, we will be able to respond by asking them to acknowledge that we have an agreed-upon WIP limit. It is likely that our system will be full and that accepting another item, however, important will break that limit. So our answer should be:

"Yes, we'd love to accept this new work, as we realize it is very important to you. Equally, you know and understand that we have an agreed-upon WIP limit. You were part of that decision, and you understand why we made it. We want to be able to process requests reliably and in a timely manner. In order to take on your request, we will need to put something else aside. Which one of the current items in progress would you prefer that we drop in order to start your new item?"

If we hadn't included our partners in the WIP-limit decision, we'd be unable to have this discussion. They would simply continue to push us. Our WIP-limited pull system would be broken and our organization would be slipping down the precipitous path back to a push system.

If we are to have a truly successful collaborative game of Kanban software development, the rules for that game must be agreed upon by consensus among all the stakeholders.

Prioritization

We also want to agree on a mechanism for queue replenishment. Typically, we are looking for an agreement to have a regular replenishment meeting and a mechanism for how new work will be selected. We can have this conversation by asking, "If we were to ask you a very simple question such as, 'What two things will you need delivered 42 days from now?' how often would you be able to meet with us to have such a discussion? We'd hope that the meeting would take no longer than 30 minutes." Because you are offering to make the meeting extremely focused, and you are asking a very direct question and suggesting that the time commitment is minimal, you will typically find that upstream partners are quite willing to be very collaborative. It's not unusual to get agreement on a weekly meeting. More often than that is common in fast-moving domains such as media, where the release cycles may be very frequent.

Delivery/Release

Now we must agree to a similar thing with downstream partners. A delivery cadence that makes sense is very specific to a domain or situation. If it's web-based software, we have to deploy to a server farm. Deployment involves copying files and perhaps upgrading a database schema and then migrating data from one version of the schema to another. This data migration will probably have its own code and it will take its own time to execute. To calculate the total deployment time, you will need to factor in how many servers, how many files to copy, how long it takes to pull systems down gracefully and reboot them, how long data takes to migrate, and so forth. Some deployments may take minutes, others hours—or even days. In other domains, we may need to manufacture physical media such as DVDs, package them in boxes, and distribute them through physical channels to distributors, dealers, retailers, or existing corporate customers. There may be other elements involved, such as printing of physical manuals or training the sales and support staff. We might need to devise a training program for these people.

For example, in 2002, I took part in the release of the first of the staged upgrades to the Sprint PCS mobile phone network. This first upgrade on the road to 3G technology was called 1xRTT. It was launched to the market as PCS Vision. The launch involved release of around 15 new handsets with a target of 16 new features that utilized the high-speed data capabilities of the new network. Sprint had a retail network across the United States that employed 17,000 people. They had a similar number of folks in call centers who took customer-care calls from users. Both the retail sales channel and the customer-care associates had to be trained to support the launch of the new service. I jokingly suggested that the best way to do this would be shut everything down for two days, fly everyone into Kansas City for a night, and rent the Kansas City Chiefs' stadium, where we could deliver a two-hour PowerPoint presentation on the big screens at either end of the stadium. This might have been the most efficient way, but it was totally unacceptable for several reasons. Our customers would hardly accept a 48-hour support outage while we trained our operators on the next generation of technology. And losing two days of sales revenue from the retail channel wouldn't have helped our annual revenue targets.

A training program was devised, and train-the-trainer education was delivered. A program for training regional retail staff was devised, as was a similar one for call centers. Trainers were sent out into the field for six weeks to train small groups of people as they came off shift. The cost of delivering the training was huge. The time commitment—six weeks—was significant, and the half-life of the training in the memories of the workforce was also about six weeks. If we missed the launch window for the new service, the training would have to be repeated, and deployment would have been delayed a minimum of six weeks further.

If your domain is like a telephone network, you know that the release cadence will be infrequent. When the transaction costs of making a release include six weeks of training, releasing any more frequently than annually is prohibitive.

The outcome you desire is the most frequent release cadence that makes sense. So start by asking, "If we give you high-quality code with minimal defects, and it comes with adequate notice, transparency into its complexity, and reliability of delivery, how often could you reasonably deploy it to production?" This will provoke some discussion around the definitions you've used, and some reassurance will be required. However, you should push for a result that maximizes business agility without over-stressing any one part of the system.

Lead Time and Classes of Service

When it comes to our conversation about lead time, it helps to have some historical data on past performance. Ideally, we want to have lead-time and engineering-task time data. In the Microsoft example from chapter 4, we knew that lead times were around 125 days for severity-1 defects and 155 days for other severities. The first thing that should strike you about this is that there are two classes of service. Severity-1 defects have historically received some form of preferential treatment. There may never have been any formality around this, but the net outcome is that severity-1 defects were being processed faster.

Knowing this may enable us to offer two different classes of service from the get-go. We may suggest to the external stakeholders that we will adopt two classes of service and have separate lead time targets for each.

Equally, we also knew from historical data that average engineering effort was 11 days and that the high end was 15 days. So we chose to suggest a 25-day lead time from selection into the input queue. There was no more science involved than that. Now, imagine the psychological effect of this. The business was used to a performance in the four to five months range and we had just offered 25 days. The difference is that we'd offered 25 days of lead time not including any initial queuing and the 155 days was a lead time that did involve queuing. Nevertheless, it sounds like a fantastic improvement. It is not surprising that the business agreed.

Other alternatives exist. You might take the historical engineering-effort data and place it in a statistical process-control chart. This will give you an upper-control (or 3-sigma) limit. You may then want to buffer that upper-limit number with a small amount for safety that will absorb external variations. If you do this, though, you should be transparent with partners and show them how you are calculating the numbers.

Another alternative would be to ask what level of responsiveness the business actually needs. This would be best done in the context of a set of classes of service. For example, if the business answers, "We need delivery in three days." You might reply, "Does everything need to be delivered in three days?" The answer is almost certainly, "No." That would give you the opportunity to ask for a definition of the types of requests that need delivered within three days. You can then create a class of service for this type of work. Then repeat the process for the remaining work. The outcome should be the stratification of work requests into several bands for which a class of service can be devised. It is likely that each of these bands will contain work that exhibits the same shape of function for cost of delay. The detail around creating classes of service and the concept of cost-of-delay functions is explained fully in chapter 11.

The lead-time target you are agreeing to for each class of service should be presented as a target rather than a commitment. You will commit to doing your best to achieve the target time and to report due-date performance against the lead-time target in the service-level agreement (SLA) for each class of service. In some situations, there may not be sufficient trust to allow agreement that lead time in the SLA is a target rather than a commitment. If you do need to agree that lead time in the SLA represents a commitment, you should buffer the target with a margin for safety. This will highlight directly that a lower level of trust results in a direct economic cost.

The exit criteria for your partner discussions is this: You have a consensus on WIP limits along the value stream; you have an agreement on prioritization coordination and the method to be used; you have a similar agreement on delivery coordination and method; and you have a definition of a set of service-level agreements that include a target lead time for each class of service.

Takeaways

❖ There are at least eight possible goals for introducing Kanban to your organization.

❖ Improve performance through process improvements introduced with minimal resistance.

❖ Deliver with high quality.

❖ Deliver a predictable lead time by controlling the quantity of work-in-progress.

❖ Give team members a better life through an improved work/life balance.

❖ Provide slack in the system by balancing demand against throughput.

❖ Provide a simple prioritization mechanism that delays commitment and keeps options open.

❖ Provide a transparent scheme for seeing improvement opportunities, thereby enabling change to a more collaborative culture that encourages continuous improvement.

❖ Strive for a process that enables predictable results, business agility, good governance, and the development of what the Software Engineering Institute calls a high-maturity organization.

❖ It's important to define your goals and to be able to articulate the benefits of introducing Kanban in order to gain a consensus agreement with other stakeholders.

❖ Follow the 12-step guide to bootstrapping a Kanban process.

❖ Kanban strikes a different bargain with external stakeholders and business owners. It is a bargain based on an assumption of a long-term relationship and a commitment to system-level performance.

❖ Including external stakeholders to form an agreement on the basic elements of the Kanban system makes them collaborators.

❖ Basic policies on WIP limits, lead-time targets, classes of service, prioritization, and delivery represent the rules of the collaborative game of software development.

❖ Involving external stakeholders as collaborators to agree on the rules of the game will enable collaborative behavior later, when the system is put under stress.

❖ PART FOUR ❖

Making Improvements

❖ **Chapter 16** ❖

Three Types of Improvement Opportunity

Chapters 6 through 15 describe how to build and operate a kanban system and adopt the Kanban approach to change management and improvement. The remainder of the book describes how to recognize opportunities for improvement, what to do about them, and how to choose between them.

Chapter 2 identifies the five core properties you'd expect to find in an organization using Kanban. The fifth property describes how models are used to identify, evaluate, and drive improvement opportunities. There are many possible models. This chapter focuses on three common models and some of their variants: the Theory of Constraints and its Five Focusing Steps; a subset of ideas from Lean Thinking that identifies wasteful activities as economic costs; and some variants that focus on understanding and reducing variability. Other models are possible. The community is already experimenting with models such as Real Option Theory and Risk Management. What follows here are examples. They are a starting point. I would encourage you to adopt them, as I know they work, and I would further encourage you to expand your thinking and look to a wider variety of models that enable and empower teams to generate improvements.

Bottlenecks, Waste Elimination, and Reduction of Variability

Each of these models for improvement has been fully explored and developed in its own body of knowledge. Each has its own school of thought on continuous improvement. With Kanban, I have chosen to synthesize all three and to provide an overview of how to recognize these improvement opportunities, and details on how to implement improvements using each model. Each of the three schools of thought on continuous improvement described below has its own group of thought leaders, its own conferences, its own canon of knowledge and experience, and its own group of followers. Your company may subscribe to one or more of these schools. Being able to show how Kanban's techniques can provide opportunities for improvement in your organization's favorite flavor may be an advantage. Knowing that you have a wide set of improvement paradigms and tools to choose from should provide greater flexibility to make change.

Those widely familiar with continuous improvement methodologies may choose to skip the rest of this chapter and move straight to chapter 17. Those who wish an overview of available methods, and some background on the literature and history, may find the remainder of this chapter valuable.

Theory of Constraints

The Theory of Constraints was developed by Eli Goldratt and first published in his business novel, *The Goal,* in 1984. Over the last 25 years, *The Goal* has gone through several revisions and the theoretical framework known as the "Five Focusing Steps" has become more obvious in more recent editions.

The Five Focusing Steps is the basis for continuous improvement in the Theory of Constraints. It is known as a POOGI (Process Of OnGoing Improvement). The Theory of Constraints (or TOC) is full of acronyms. Strangely, Five Focusing Steps is the exception. It is not abbreviated as "FFS."

In the 1990s, the Theory of Constraints evolved a method for root-cause analysis and change management known as the Thinking Processes (or TP). The reason for this development was the discovery among the TOC consulting community that their constraint on achieving improvement with clients was change management and resistance to change.

It seemed that the Five Focusing Steps only appeared to work well for flow problems and that many workplace challenges did not fit neatly into the flow paradigm. So TP evolved. The professional qualification and training program for TOC consultants changed from a class on the use of the Five Focusing Steps and its applications, such as Drum-Buffer-Rope, to a class on TP. Hence, many in the TOC community, when referring to TOC, are in fact referring to TP and not the Five Focusing Steps. In my observation while attending TOC conferences, use of the Five Focusing Steps among the TOC community has become somewhat of a lost art.

The TOC community has, from what I've seen, tended to accept paradigms as they are established rather than challenge them. Hence, the TOC solution for project management, Critical Chain, evolved around the incumbent project-management paradigm of the triple constraint (scope, budget, and schedule) and the dependency-graph model for scheduling the tasks in a project. No one challenged the incumbent model. No one, until I published my first book, *Agile Management for Software Engineering,* challenged the project-management paradigm and suggested it was better to model projects as a value-stream and flow problem and apply the Five Focusing Steps. By doing this, it was then possible to use the whole Lean body of knowledge, based around flow, and synthesize it with the Five Focusing Steps' focus on bottlenecks. The synthesis of TOC with Lean enabled improvements in project and organizational performance and laid the foundation for the emergence of Kanban.

I have argued that any process or workflow that involves division of labor can be defined as a value stream; and any value stream can be observed to have flow. Lean and the Toyota Production System are essentially built around this assumption. If any value stream has flow, then the Five Focusing Steps can be applied to it. Hence, the Five Focusing Steps is a perfectly satisfactory POOGI, and TP is not required unless you are using it as a change-management tool. I personally have not developed an affinity with TP. My preferred change-management tool is Kanban, as this text lays out.

Five Focusing Steps

The Five Focusing Steps is a simple formula for a process of ongoing improvement. It states:

1. Identify the constraint.
2. Decide how to exploit the constraint.
3. Subordinate everything else in the system to the decision made in Step 2.
4. Elevate the constraint.
5. Avoid inertia; identify the next constraint and return to step 2.

Step 1 is asking us to find a bottleneck in our value stream.

Step 2 asks us to identify the potential throughput of that bottleneck and compare that to what is actually happening. As you will see, the bottleneck is rarely or never working at its full capacity. So ask, "What would it take to get the full potential out of our bottleneck? What would we need to change to make that happen?" This is the "decide" part of step 2.

Step 3 asks us to make whatever changes are necessary to implement the ideas from step 2. This may involve making additional changes elsewhere in the value stream in order to get the maximum capacity from the bottleneck. This action of maximizing the bottleneck's capability is known as "exploiting the bottleneck."

Step 4 suggests that if the bottleneck is operating at its full capability and is still not producing enough throughput, its capability needs to be enhanced in order to increase throughput. Step 4 asks us to implement an improvement to enhance capability and increase throughput sufficiently so that the current bottleneck is relieved and the system constraint moves elsewhere in the value stream.

Step 5 requires that we give the changes time to stabilize and then identify the new bottleneck in the value stream and repeat the process. The result is a system of continuous improvement in which throughput is always increasing.

If the Five Focusing Steps is institutionalized properly, a culture of continuous improvement will have been achieved throughout the organization.

Chapter 17 explains how to identify and manage bottlenecks using the Five Focusing Steps.

Lean, TPS, and Waste Reduction

Lean emerged in the early 1990s after the seminal text, *The Machine That Changed the World,* by Womack, Jones, and Daniels, described, from an outsider's empirical observation, how the Toyota Production System (TPS) worked. The early literature on Lean had some flaws. It failed to identify the management of variability that is inherent to TPS and that was learned and adapted from Deming's System of Profound Knowledge. Lean also fell victim to misinterpretation and over-simplification. Many Lean consultants jumped on the concept of Waste Reduction (or elimination) and taught Lean as purely a waste-elimination exercise. In this anti-pattern of Lean, all work activities are classified as value-added or non-value-added. The non-value-added, wasteful activities, are further sub-classified into necessary and unnecessary waste. The unnecessary activities are eliminated and the necessary are reduced. Although this is a valid use of Lean tools for improvement, it tends to sub-optimize the outcome for cost reduction and leaves value on the table by not embracing the Lean ideas of Value, Value Stream, and Flow.

Kanban enables all aspects of Lean thinking and provides the tools to optimize an outcome for value through a focus on flow management as well as waste reduction.

Chapter 18 explains how to identify wasteful activities and what to do about them.

Deming and Six Sigma

W. Edwards Deming is generally thought of as one of the three fathers of the Quality Assurance movement of the twentieth century. However, his contribution was considerably greater. He evolved the use of Statistical Process Control (SPC) and developed it into a management technique he called the System of Profound Knowledge. His system was intended to prevent managers from making poor-quality decisions and replacing them with statistically sound, objective, often counter-intuitive, better decisions. Deming is occasionally mentioned as perhaps the most important management scientist of the twentieth century, and in my opinion, this is greatly deserved. His contributions extended from SPC through Quality Assurance to Management Science.

Deming had a significant influence on Japanese management philosophy around the middle of the twentieth century, and his work relating to SPC and the System of Profound Knowledge is a key pillar of TPS.

While some highly mature Kanban teams, for example, at investment bank BNP Parisbas in London, have adopted SPC, SPC is beyond the scope of this book and will be addressed in a future text on advanced Kanban techniques.

However, the principles of understanding the variation in systems and work tasks that underpin SPC are very useful. Deming's predecessor, Walter Shewhart, classified variability in task performance into two categories, chance cause and assignable cause. Deming later renamed these to common cause and special cause, and in the second edition of *The New Economics,* he admitted that this was for "largely pedagogical reasons." There was no specific innovation in changing the terms. Understanding variation and how it impacts performance, and developing the capability to classify it into the two categories are necessary management skills. Learning appropriate management actions to take based on the type of variation is core to a program of continuous improvement. Both Lean and Theory of Constraints rely heavily on an understanding of variation in order to enable improvement, even if those improvements are cast as bottleneck management or waste reduction.

Chapter 19 explains how to recognize common and special cause variations and suggests ideas for appropriate management action. Chapter 20 further elaborates on this; it describes how to build an issue-management capability that responds to assignable-cause variations with the goal of eliminating such issues as quickly as possible in order to maintain flow and maximize value delivery. Note: Without a knowledge of and focus on managing variability, a focus on flow will be ineffective. Lean without Deming's ideas is Lean without an understanding of variation, and, by implication, is Lean without a focus on maintaining flow. Given that the early Lean literature did not include an understanding of variation nor any references to Deming's System of Profound

Knowledge, it is easy to understand the root cause of the anti-pattern of teaching Lean as a process of waste reduction only.

While Deming's ideas were embedded into TPS in Japan at the shop-floor level, where SPC and the System of Profound Knowledge were employed to identify local improvement opportunities, another body of knowledge developed in the U.S. based around Deming's ideas. Six Sigma started at Motorola, but really came of age when it was adopted at GE under Jack Welch's leadership.

Six Sigma employs SPC to identify common and special cause variation and uses a process similar to that described by Deming to eliminate special cause variations at their root cause and prevent them from recurring; and in addition, to reduce common cause variation and make a process, workflow, or system more predictable.

Unlike TPS, which is all about shop-floor initiatives run by empowered workers implementing small kaizen events by the hundreds and thousands, Six Sigma has developed into a much more low-trust, command-and-control method that tends to involve far fewer improvement opportunities, generally implemented at a more strategic level, and run as specific projects in their own right. The project leader carries the title Black Belt and has generally had years of training in the methodology to earn his status. Because Kanban embraces the ideas of Deming and provides the instrumentation and transparency to see variability and its effect, Kanban can be used to enable either a kaizen-style improvement program or a Six Sigma–style improvement program.

Fitting Kanban to Your Company Culture

If your company is a Six Sigma company, Kanban can help you run Six Sigma initiatives in the software, systems, product development, or IT organization. If your company is a Lean company, Kanban is a natural fit. It can enable an entire Lean initiative in your software, systems, product development, or IT organization. If you company subscribes to and uses the Theory of Constraints, Kanban can enable an entire constraints-management (bottleneck-removal) program in your software, systems, product development, or IT organization. However, you might need to recast the pull system implementation as a Drum-Buffer-Rope implementation rather than refer to is as kanban system. Because Kanban developed from an earlier Drum-Buffer-Rope implementation, I know this will work. However, discussion of the specifics of how to model the value stream and set WIP limits for the Buffer and Rope are beyond the scope of this text.

Takeaways

❖ Kanban requires that models are used to identify improvement opportunities.

❖ Kanban supports at least three types of continuous improvement methods: Constraint Management (bottleneck removal), Waste Reduction, Variability Management (as well as SPC and the System of Profound Knowledge).

❖ Kanban enables the identification of bottlenecks and a full implementation of the Five Focusing Steps from the Theory of Constraints.

❖ Kanban enables visualization of wasteful activities and can be used to enable a full Lean initiative within the software, system, product development, or IT organization.

❖ Kanban provides the instrumentation for use of W. Edwards Deming's Theory of Profound Knowledge and Statistical Process Control. It can be used to drive a kaizen initiative or a Six Sigma initiative.

❖   Chapter 17   ❖

Bottlenecks and Non-Instant Availability

Washington SR-520 is the freeway that links Seattle with its north-eastern suburbs of Kirkland and Redmond. It provides the main commuter artery for suburban dwellers who work in the city center and for employees of Microsoft and the other high-technology firms based in those suburbs, such as AT&T, Honeywell, and Nintendo, who live in the city and commute in the opposite direction each weekday. For a total of eight hours each day, the road is a severe traffic bottleneck in both directions. If you stand on the bridge that crosses the freeway at N. E. Seventy-sixth Street in the small suburb of Medina (just up the street from Bill Gates's estate on the shores of Lake Washington), in late after-noon and look eastward, you will see the westward, city-bound traffic backing up and crawling slowly up the hill from Bellevue before merging down to two lanes to cross the floating bridge into Seattle. The speed of traffic coming up the hill is about 10 miles per hour and the flow is ragged with vehicles constantly slowing and stopping. If you cross the street and look westward toward Seattle's downtown skyscrapers, the Space Needle, and the Olympic Mountains in the far distance, you'll see the traffic moving smoothly away from you at almost 50 miles per hour. What magic is happening right beneath your feet that the traffic speed changes so dramatically and its flow transforms from ragged to smooth?

Just before the bridge, the road narrows from three lanes to two before crossing the lake on the pontoon bridge. The rightmost lane of the freeway is a high-occupancy vehicle (HOV) lane that requires vehicles to have two or more passengers. It is frequented by the many public-service buses that shuttle commuters to and from the city and some private cars. The action of these vehicles merging into the other traffic is enough to cause disruption and a slowdown of, and backup in, traffic. In the several miles that precede the bridge, several other roads merge onto the freeway, adding additional traffic volume to what is already a busy road at peak times. The net effect is ragged flow and very slow speeds.

In traffic safety, the planners worry about the distance between cars. Ideally, they want enough distance for cars to react to changes and to stop safely if necessary. This distance is related to speed and reaction time. The legally advised "distance" between vehicles is recommended at two seconds. In Lean language this is the ideal takt time between vehicles. Hence, if we have two lanes and two seconds between vehicles the maximum throughput of the road is 30 vehicles per lane per minute or 60 vehicles per minute. This is true regardless of the speed of the vehicles. These rules break down at extreme limits for very slow speeds and for super-excessive speeds—those well in excess of the 50-miles-per-hour limit enforced on SR-520. For practical purposes the throughput (referred to confusingly as capacity in traffic management) is 3,600 vehicles per hour.

However, as you stand on the bridge and count the number of cars passing under it on a typical afternoon around 5 P.M., you'll observe that fewer than 10 cars per minute are crossing onto the floating bridge toward Seattle. Despite the heavy demand, the road is operating at less than one-fifth of its throughput potential! Why?

The pontoon bridge over Lake Washington is a bottleneck. We all intuitively understand this concept. The width of a bottle's neck controls the flow of liquid into and out of the bottle. We can pour quickly from a wide neck, but often with greater risk of spillage. With a narrow neck, the flow is slower but it can be more precise. Bottlenecks restrict our potential for throughput, in this example, from 60 cars per minute, or 3,600 per hour, to fewer than 10 cars per minute, or 600 cars per hour.

In general, a bottleneck in a process flow is anywhere that a backlog of work builds up waiting to be processed. In the example of SR-520, that backlog is a queue of vehicles occasionally backed up to Overlake, seven miles east. In software development, it can be any backlog of unstarted work or work-in-progress: requirements waiting for analysis; analyzed work waiting for design, development, and testing; tested work waiting for deployment; and so forth.

As discussed, SR-520 delivers only about 20 percent of its potential at peak times when it is needed most. For a full explanation of this, we need to understand both how to fully exploit a bottleneck's potential and the effect that variability has upon that potential. These concepts are explained here in chapter 17 and later in chapter 19.

Capacity-Constrained Resources

SR-520 at the N. E. Seventy-sixth Street Bridge is a capacity-constrained bottleneck. Its capacity is 60 cars per minute in two lanes. Leading up to this, the road is three lanes wide, so traffic is forced to merge together in order to cross the lake on the aging pontoon bridge that was designed 50 years ago with only two lanes. At the time, that was plenty of capacity and the bridge was not a bottleneck. The eastern suburbs were small villages, and commuting across the water was rare—and in those days, only toward the city and not in the reverse as is common today.

Elevation Actions

In this respect, as a capacity-constrained bottleneck, SR-520 may be similar to a user-experience designer on a software team who is responsible for designing all the screens and dialogs with which the user interacts. She works flat out but still her throughput is insufficient to meet the demand placed on her by the project. The natural reaction of most managers in this situation is to hire another person to help. In Eli Goldratt's Theory of Constraints, this is known as "elevating the constraint"—adding capacity so that the bottleneck is removed.

In our example of SR-520, this would be the equivalent of replacing the floating pontoon bridge across Lake Washington with a new bridge that features three lanes of traffic each way. To keep all things equal, it should be a bridge comprised of one HOV lane and a bicycle lane, as well as two lanes open to all traffic. This is, in fact, the course of action that the Washington State Department of Transportation is pursuing. The bridge will cost many hundreds of millions of dollars and will take a decade to implement. At the time of this writing, construction has not started.

It turns out that elevating a capacity-constrained resource ought to be the last resort. Increasing the capacity of a bottleneck costs both time and money. If, for example, we have to hire another user-experience designer, we need to find the budget to pay this new person as well as the budget to fund the hiring process, which would include any fees we may pay agents for referrals. We will slow the progress on our current project while we review résumés and interview candidates. Our most precious resource, our capacity-constrained user-experience designer, will be asked to take time out from real project work to read résumés, select candidates, and then interview them. As a result, her capacity to complete designs is reduced, as is the potential throughput for our whole project. This is partly why Fred Brooks's "Law" states that adding people to a late project only makes it later. Although Brooks's observation was anecdotal, and we can now make a much more scientific explanation of this phenomenon, the software industry has understood, for at least the last 35 years, the concept that hiring more people slows a project down.

Exploitation/Protection Actions

Rather than jump immediately to elevation, and spend time and money while slowing things down, it is better to first find ways to fully exploit the capacity of the bottleneck resource. For example, SR-520 is observed to have a throughput only 20 percent of its potential at peak times. What actions might be taken to improve that throughput? Let's dream for a moment. If the throughput of the road at rush hour were achieving its potential of 3,600 vehicles per hour, would it be necessary to replace the existing bridge with a new one? Would journey times be sufficiently short that Washington State taxpayers (like the author) might prefer to spend their tax dollars on some other, more important and pressing matter? Such as more books in local schools? Perhaps!

So how would you go about exploiting the true potential of the road? The source of the problem actually lies with the humans driving the vehicles. Their reaction times and the actions they take are highly variable. As cars merge from the HOV lane, vehicles in the center lane need to slow down and make space for a merging car. Some drivers react more slowly than others; some stand on the brakes more vigorously than others, and the net effect is that traffic backs up unpredictably. Some drivers, disturbed by the fluctuation in the lane ahead and the reduced speed in comparison to the neighboring left lane, decide to switch and jump across, merging into the left lane. The same effect is then repeated. All the vehicles have slowed, but the speed does not really affect throughput. It's the gap between the vehicles that is most important. What is desired is a smooth flow of traffic with a two-second gap between vehicles.* However, the human element means that vehicles do not slow down or accelerate smoothly, and the gaps accordion. The reaction time of the individuals to press the accelerator and brake pedals, and the reaction times of the engines and transmissions and gear boxes in the vehicles means that gaps continue to widen as traffic builds up. Variability in the system has a huge impact on throughput.

Fixing this problem for SR-520 takes us into fantasyland in terms of vehicle control, though some German manufacturers have experimented with such systems. Systems that use radar or lasers to judge the distance between vehicles and keep traffic moving in a smooth convoy can remove the variability that occurs on SR-520. Such systems have the ability to slow whole chains of vehicles smoothly while maintaining the gap between them. As a result, the throughput of traffic remains high. However, eliminating variability

---

*   In some parts of California, 1.4 seconds is observed as normal, though not ideal from a safety perspective.

from private vehicles driven by the occupants has its limits. If you want low-variability transportation, you'd have to chain the passenger cars together and put them on rails. That's fundamentally why mass–rapid rail transport is more effective than automobiles at moving large quantities of people quickly.

The good news is that in our office, our capacity-constrained resources are affected by variability that we *can* do something about. We've talked a lot in this book about coordination activities and transaction costs of doing value-added work. If we have a capacity-constrained user-experience designer, we can seek to keep that individual busy working on value-added work by minimizing the non–value-added (wasteful) activities required of that person.

For example, I had a capacity-constrained test team in 2003. To maximize the exploitation of their capacity, I looked for other, slack resources and found them with the business analysts and a project manager. The test team was relieved of bureaucratic activities such as time-sheet completion. They were also relieved of planning future projects. We allowed analysts to develop test plans for future iterations and projects while the testers busied themselves performing tests on the current work-in-progress.

Another, and better, approach, one that I did not consider at the time, would have been to devise a risk profile for requirements that must be tested by only the professional test team. Requirements not meeting the criteria could then be tested by people from other functional areas playing a dilettante, or amateur, testing role; for example, the business analysts. This "bifurcation" technique, using a risk profile, is a very good way to optimize the utilization of a bottleneck while continuing to manage risk on the project.

A long-term fix may have been to invest heavily in test automation. The key word in the last sentence is "invest." If you find yourself saying "invest," you are generally talking about an elevation action. Adding resources is not the only way to elevate capacity. Automation is a good and natural strategy for elevation. The Agile software development community has done a lot to encourage test automation in the last decade. As a general rule, consider automation as an elevation strategy. However, a wonderful side effect of automation is that it is also reduces variability: Repeatable tasks and activities are repeated with digital accuracy. So automation reduces variability in the process and may help to improve exploitation of capacity at another bottleneck.

The next way to ensure maximum exploitation of our capacity-constrained user-experience designer is to ensure that she is always making progress on current work. If the user-experience designer reports that she is blocked for some external reason, the project manager and, if necessary, the whole team should swarm on the issue to get it resolved. A strong organizational capability at issue identification, escalation, and resolution is essential for effective exploitation of capacity-constrained bottlenecks.

If there are several issues blocking current work, then the issues impeding the capacity-constrained resource, in this case our user-experience designer, should get highest priority. For effective, high-performance issue management, it is therefore necessary to know the location of the capacity-constrained resource and to give it priority when required.

The transparency in the Kanban system will help to raise awareness of both the location of the capacity-constrained (bottleneck) resources and the impact of any issues impeding flow at that point in the system. With everyone on the project aware of the system-level impact of an impediment on the bottleneck, the team will gladly swarm on a problem to resolve it. Senior management and external stakeholders with a vested interest in a release arriving on time also will give of their time more freely when they understand the value of that time and the impact that swift resolution of an issue will provide.

Hence, developing an organizational capability of transparently tracking and reporting on projects using a kanban system is critical to improving performance. Transparency leads to visibility of both bottlenecks and impediments and, consequently, to improved exploitation of available capacity to do valuable work through a team focus on maintaining flow.

One more technique that is commonly used to ensure maximum exploitation of a capacity-constrained resource is to ensure that the resource is never idle. It would be a terrible waste if the capacity-constrained resource was left without work to do because of an unexpected problem upstream; for example, a requirements analyst takes several weeks off work due to a family medical issue. Suddenly the constraint is moved. Or perhaps a large section of the requirements is recalled by the business, which has made a strategic change. While the team waits for new requirements to be developed, the user-experience designer is idle. What if the upstream activities are highly variable in nature? This is common with requirements solicitation and development. Hence, the arrival rate of work to be done may be irregular. There could be many reasons that the capacity-constrained resource may become idle due to a temporary lack of work. The most common way to avoid such idle time is to protect the bottleneck resource with a buffer of work. The buffer is intended to absorb the variability in the arrival rate of new work queuing, in this example, for our user-experience designer. Buffering adds total WIP to our system. From a Lean perspective, adding a buffer of work adds waste, and it increases lead times. However, the throughput advantage provided by ensuring a steady flow of work through our capacity-constrained resource is usually a better trade. You will get more work done despite the slightly longer lead time and the slightly greater total work-in-progress.

Using buffers to ensure a bottleneck resource is protected from idle time is often referred to as "protecting the bottleneck" or a protection action. Before considering elevating a bottleneck you should seek to maximize exploitation and protection to ensure that the available capacity is utilized as much as possible.

Our traffic-management example with SR-520, where the actual throughput was less than 20 percent of potential, turns out to be quite common with knowledge-work problems such as requirements analysis and software development. It is often possible to see improvements of up to four times in delivery rate by exploiting a bottleneck.

In the example from Microsoft in chapter 4, a two-and-one-half-times improvement was achieved through better exploitation and protection that removed variability from the system without spending any money or elevating the bottleneck.

Subordination Actions

Once you've decided how to exploit and protect a capacity-constrained resource, you may need to take action—to subordinate other things in the system—to make your exploitation scheme work effectively.

Let us revisit our fantasy traffic system in a future world. Here, we decide not to build a new floating bridge across Lake Washington; instead we decide to fit all vehicles traveling on SR-520 at peak times with a new velocity-management system that uses radar and wireless communications to regulate the speed of traffic on a seven-mile stretch of freeway. This new system would act like cruise control and override the manual use of the accelerator and brake pedals. Citizens would be incentivized to fit the system to their vehicles with tax breaks. Once enough cars have the system, it would be switched on and cars without it would have to find an alternative route or choose to cross outside of peak times. The result would be smoother traffic flow and greater exploitation of capacity at the bottleneck. My guess is that such a system, if it could be effective, would reclaim about 50 percent of the lost capacity. Or put another way, it would increase throughput across the SR-520 in peak times by about two and one-half times.

So what have we done in this example? We have subordinated the driver's right to affect and control their own speed in pursuit of the greater common goal of faster journey times facilitated by greater throughput across the bridge. This is the essence of a subordination action. Something else will need to change in order to improve the exploitation of the bottleneck.

For those knowledgeable about the Theory of Constraints it is often counter-intuitive to realize that the changes required to improve performance in a bottleneck are usually not made at the bottleneck. While reviewing the manuscript for my first book[20] a now well-known member of the Agile software development community suggested that using the Theory of Constraints as an approach to improvement would lead to everyone on the team wanting to be part of the bottleneck resource because they would get all the management attention. This is an easy mistake to make. Counter-intuitively, most bottleneck management happens away from the bottleneck. Many of the changes focus on reducing failure load to the bottleneck in order to maximize its throughput. As a general rule, expect to maximize exploitation of bottleneck capacity, and hence maximize throughput, and as a result, minimize the delivery time on your project by taking actions all over the value stream, and most likely not at the bottleneck itself.

Non-Instant Availability Resources

Non-instant availability resources are not, strictly speaking, bottlenecks. However, by and large they look and feel like bottlenecks, and the actions we might take to compensate for them are similar in nature to those for a bottleneck. Anyone who has ever driven a car and stopped at a traffic light understands the concept of non-instant availability. While stopped at a red light, the car cannot flow down the road. The lack of flow is not caused by capacity constraints on the road but by a policy that allows cars traveling on another road the right to cross the road on which our car is traveling.

A better example, and sticking to our theme in this chapter of transportation in Washington State, would be the ferry system that operates across Puget Sound, linking the Kitsap and Olympic Peninsulas with the mainland around the city of Seattle. There are three ferry crossings, two that leave from Seattle crossing to Bremerton and Bainbridge, and my favorite, the SR-104 crossing between Edmonds on the east side to Kingston on the west. On a map, the ferry route is actually shown as part of the SR-104 road. It is often marked as "toll," rather than explicitly saying "you have to get on a boat here ;-)." The transportation people think of the ferry as a non-instant availability road.

When you show up for the ferry, you pay some money and are asked to wait in a holding area. A typical wait time is about 30 minutes, as the ferry takes 30 minutes or so to cross Puget Sound, and there is a 10- to 15-minute period to unload the vehicles, and another similar period to load all the new vehicles before setting sail. Usually the ferry company is operating two boats, so boats sail every 50 minutes or so. At peak times they may operate three boats on the route, shrinking the wait time between sailings to around 35 minutes.

Most of the time, the ferry sails close to full, but the system is not capacity constrained. The fact that cars build up in a holding area—a buffer—and are loaded onto the ferry for sailing (batch transferred) does not indicate a capacity-constrained resource. It does, however, indicate a non-instant availability resource. Ferries sail only once or twice an hour, with a capacity of around 220 cars per sailing.

At peak times, such as Friday afternoon, the ferry system does become capacity constrained. When this happens, the arrival rate of cars wishing to cross exceeds the capacity to transport them. The capacity is roughly 300 cars per hour. Cars begin to back up, queuing outside the holding area, before the toll booth. During these peak demand times, you can often see vehicles snaking back two miles through Edmonds or Kingston. There is little that can be done; cars just have to wait. It isn't easy to elevate the constraint by bringing on another ferry. The timetable and schedule of ferry sailings is designed to provide a reasonable level of service, for a reasonable amount of the time. To always have excess capacity would be excessively expensive on the state taxpayers who subsidize the ferry service.

Moving back to software development and knowledge work, non-instant availability tends to be a problem with shared resources or people who are asked to perform a lot of multi-tasking. As we all know, there really is no such thing as multi-tasking in the office; what we do is frequent task switching. If we are asked to work on three things simultaneously, we work on the first thing for a while, then switch to the second, then to the third. If someone is waiting for us to finish the first thing while we are working on the second or third, then we would appear to be non-instantly available from that person's (and the first task's) perspective.

One example of non-instant availability that I observed occurred with a build engineer. The company had a policy that only configuration-management team personnel were allowed to build code and push it into the test environment. This policy was a specific risk-management strategy based on historic experience that developers were often careless and would build code that would break the test environment. The test environment was often being shared among several projects, and hence the impact of a bad build was significant. The technology department did not have a good program-level coordination capability, and the likelihood that one team and one project was working in an area of the aggregate IT systems that might affect that of another project was quite high. The coordination function of knowing what was happening at a technical code level between and across projects was given to the configuration management department. These professionals were known as build engineers. The build engineer was responsible for knowing the impact of a set of changes in a given software build and for avoiding breaking the test environment so that the flow of all projects was not affected by an outage in the test environment.

Generally, a project had a build engineer assigned from the shared pool of configuration-management team resources. However, the demand from a single project for code builds into test was not sufficient to keep a build engineer busy for a full day. In fact, it generally wasn't enough to keep him or her busy for more than an hour or two each day. Hence, build engineers were asked to multi-task. They were either assigned to several projects or were assigned other duties.

Take the example of Doug Burros at Corbis; he was assigned as the build engineer for the sustaining engineering activity. He was also assigned two other duties: He had responsibilities for building out new environments and also for maintaining existing environments. He was the configuration-management engineer with full responsibility to keep the configurations current. This included applying operating system and database server patches and upgrades, middleware patches and upgrades, system configuration and network topology, and so forth. He allocated about one hour per day to perform the build engineering duty. Typically, this would be in the morning, from approximately 10:00 to 11:00. If developers found themselves at 3:00 P.M. requiring a test build, typically they would have to wait until the next day. The build engineer was non-instantly available. Work would block and, as sustaining engineering was operated using a kanban system, work would quickly back up along the whole value stream, causing idle time for many other team members.

The actions taken in response to non-instant availability problems with flow are remarkably similar to those for a capacity-constrained resource.

Exploitation/Protection Actions

The first thing was to recognize that Doug was a non-instant availability resource and to observe the impact that this was having. Work was backing up when he wasn't available because the kanban limits were tightly defined. Because Doug was a source of variability in flow, the correct course of action was to place a buffer of work in front of Doug. The trick was to make this buffer big enough to allow flow to continue without making it so large that Doug would become a capacity-constrained resource. I had a discussion with him about the nature of the build activity. It turned out that he could reasonably build up to seven items in his one hour per day of availability. So we created a buffer with a kanban limit of seven. We introduced this to the value stream and the card wall by introducing a new column called Build Ready. We had actually increased the potential total WIP in the system by around 20 percent, but it worked. Although builds were not instantly available, the upstream activities were able to keep flowing during the day. The result was a significant boost in throughput and shorter lead times, despite the increase in WIP. Another option, which, again, we did not think of at the time, would have been to ask Doug to work two half-hours, rather than one full hour. These two half-hours could have been split across the day, one in the morning, and one in the afternoon. This would have eased the flow. The effect would have been to reduce the pressure to buffer the non-instant availability. Perhaps the buffer size could have been reduced to two or three. This would have had an impact of only 10 percent on total WIP and would have resulted in shorter lead times through the system.

As a general rule, when encountering non-instant availability problems, think about how to improve the availability. The ultimate is to turn a non-instant availability problem into an instantly available resource.

Subordination Actions

As discussed earlier, subordination actions generally involve making policy changes across the value stream to maximize the exploitation of the bottleneck. What options were available as subordination actions with our non-instantly available build engineer?

The first thing was to examine the policy of asking Doug to perform three different functions. Was it the best choice? I discussed this with his manager. It seemed that on her team, the engineers liked and needed diversity of work to keep it interesting. Also, by asking team members to perform system build out, system maintenance, and build engineering, it maintained a generalist skill set across all team members and kept the resource pool flexible. This provided the manager with many more options and avoided the potential for capacity-constrained resource bottlenecks due to a high degree of specialization. Generalization was also appealing to team members from a career and résumé perspective. They didn't want to become too narrowly skilled. So asking the team to work in only one area such as build engineering was not desirable.

Another option might have been to abandon the idea of multitasking and dedicate Doug to the sustaining engineering team effort. This would have provided him with a lot of idle time. He would be sitting around waiting for work, like a firefighter in the firehouse, sitting around waiting for a call that there is a fire to put out. Keeping Doug on constant standby would certainly cure the flow problems but was it a reasonable choice?

Budgets were tight and adding people to the configuration management team to handle the system build out and maintenance that Doug was doing would be expensive and perhaps impossible. I would need to ask my boss for budget to get another person because I wanted to keep someone idle most of the time. Was this a good risk-management tradeoff?

To decide this we need to look at the cost of delay for the sustaining engineering effort and compare the cost of another staff member to the cost of other alternatives to maintaining flow. The reality was that very few items in the sustaining engineering queue had a strategically significant cost of delay. So the idea that we'd keep someone idle, waiting for work, in order to optimize flow was not a viable alternative. Clearly, the exploitation action of adding a buffer of work to maintain flow was a cheaper and better alternative.

However, the discussion of what to do about Doug's lack of instant availability did create a debate on the team about the policy of having only build engineers perform this work. We discussed the option of ending the policy and allowing developers to build code and push it to the test environment. It was rejected because the organization had no viable alternative method of coordinating the technical risks across projects. One option, providing a dedicated test environment for that project, was rejected for cost reasons and wasn't a practical viable alternative in the short- or medium term. Everyone continued to see the value in the build engineering function and the configuration management team.

Elevation Actions

However, buffering and adding WIP to solve the problem felt like a Band-Aid over the wound. It felt like a workaround. And you could view it that way. It was a tactical fix—an effective tactical fix, but nevertheless a tactical fix—that carried a cost. Because the Kanban system had exposed the non-instant availability bottleneck and allowed the team to have a full debate around its cause and the possible fixes, the discussion inevitably came around to whether having a human build the code was the right answer. Would it be possible to automate this build process? The answer was "Yes," though the investment was heavy. Considerable development of capability in configuration management and cross-project coordination would be needed. In addition, some specialists in automation would have to be hired for a period of time to create the system to make it work.

It took around six months in total elapsed time, and two contractors for eight weeks. The total financial cost was around $60,000. However, the end result was that Doug was no longer required for builds, and builds were instantly available when developers needed them. At this point, it was possible to eliminate the buffer and reduce the system WIP. That in turn resulted in a slight reduction in lead time.

Automation was ultimately the route to elevation of the non-instant availability bottleneck. Adding capacity, that is, hiring another engineer, was not a good choice.

One other path involving automation was also pursued—virtualization of environments. Virtualization was already commonplace in our industry; however, at that time our test environments were still physical. Virtualization was not an organizational capability. By taking time to make it so, the test environment could be easily configured and restored. This would reduce the impact of a build breaking the environment; in risk management this is a mitigation strategy. It would also enable dedicated environments—reducing or eliminating the risk of a build breaking another project's configuration.

So buffering was used as a short-term, tactical-exploitation strategy while automation was pursued as a long-term elevation strategy.

And what about our Edmonds-to-Kingston ferry example? How might that be elevated? Well, the State of Washington is currently considering two options. One would be to replace the current, aging fleet of ferries with a newer set of larger, more efficient boats. However, Washington has a lot of experience with floating bridges. There are two across Lake Washington, including the one on SR-520, which is apparently the world's longest such bridge, and another across the Hood Canal on SR-104. What is now under consideration is the possibility of building a new, record-breaking floating bridge across Puget Sound as part of SR-104 and replacing the ferry service entirely. The planned bridge would not only solve the problem of capacity constraints at peak times, it would also solve the non-instant availability issues that hamper all ferry services as an option for traffic flow. Such a bridge would open up the Kitsap and Olympic Peninsulas for faster economic growth. Perhaps 50 years from now someone else will be writing a book discussing how the SR-104 floating bridge across the Puget Sound is now a bottleneck and capacity-constrained resource during peak commuting hours?

Takeaways

❖ Bottlenecks constrain and limit flow of work.

❖ Bottlenecks come in two varieties: capacity constrained—unable to do more work; and non-instant availability—limited capacity due to limited (but usually predictable) availability.

❖ We manage bottlenecks using the Five Focusing Steps from the Theory of Constraints.

❖ Increasing the capacity at a bottleneck is known as elevation.

❖ The actions taken to elevate a capacity-constrained resource will typically be different from the actions taken to elevate a non-instant availability resource.

❖ Elevation may involve adding resources, or automation, or policy changes that make a previously non-instantly available resource instantly available.

❖ Elevation actions typically cost money and take time to implement. Elevation actions are often considered strategic investments in process improvement.

❖ Often, bottlenecks in the process are performing well below their potential capacity—below the theoretical capacity constraint.

❖ Throughput at a bottleneck can be improved up to the limit of the theoretical capacity constraint by using exploitation and protection actions.

❖ Typically, protection involves adding a buffer of WIP in front of the bottleneck. This is true for capacity-constrained resources or non-instant availability resources.

❖ Exploitation actions typically involve policy changes that control the work done by the bottleneck resource.

❖ Classes of service can be used as exploitation actions.

❖ Subordination actions are actions taken elsewhere in the value stream to enable the desired exploitation or protection actions. Subordination actions are typically policy changes.

❖ Exploitation, protection, and subordination actions are often easy and cheap to implement, as they primarily involve policy changes. Hence, maximizing the throughput of a bottleneck by fully exploiting it can be viewed as tactical process improvement.

❖ Despite the tactical nature of exploitation of a bottleneck, the gains can often be dramatic. Two-and-one-half- to five-times improvements in throughput, with consequent drops in lead time, can often be achieved at little to no cost over a short period of months.

❖ Exploitation should always be pursued first, before elevation is attempted.

❖ It is not unusual for a tactical set of exploitation and subordination actions to be implemented while a plan for a strategic change to elevate a constraint is implemented over a longer period of time.

❖ Chapter 18 ❖

An Economic Model for Lean

Waste (or *muda* in Japanese) is the metaphor used in Lean (and the Toyota Production System) for activities that do not add value to the end product. The metaphor of waste has been proving problematic with knowledge workers. Often, it is hard to accept as waste tasks or activities that are costs or overheads but that are necessary or essential to completing value-added work; for example, daily stand-up meetings are essential to coordinating most teams. These meetings do not directly add value to the end product, so, technically, they are "waste," but this has been hard to accept for many Agile development practitioners. Rather than have people wrapped up in arguments about what is or is not waste, I have concluded that it would be better to find an alternative paradigm and alternative language that is less confusing or emotionally evocative.

Redefining "Waste"

Following the lead of writers such as Donald G. Reinertsen, I have adopted the use of the language of economics and refer to these "wasteful" activities as costs. I classify costs into three main abstract categories: Transaction Costs; Coordination Costs; and Failure Load. Figure 18.1 illustrates this.

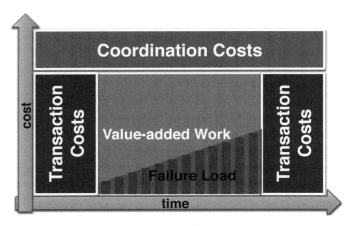

Figure 18.1 The Economic Cost model for Lean software development

Figure 18.1 shows that over time there are a number of value-added activities for an iteration or project. Surrounding those activities are transaction and coordination costs. The capacity for value-added activities can be displaced by work that can be considered failure load, that is, work that is either rework or demand placed on the system because of a previously poor implementation.

Each of these costs is discussed in detail in the following sections. I will describe them using a simple, real-world example—the activity of painting the fence at my home in Seattle with preservative wood stain.

Transaction Costs

The fence has 21 sections. Customer value is delivered when a section of fence is painted with the wood stain. Full value is delivered when all 21 sections are stained on both sides.

Before I could start the job, I first had to procure all the materials. This involved a trip to Home Depot. There was also some preparation work required on the fence: some repairs, some sanding, and trimming plants and bushes to allow access for painting. None of these activi-

ties could be described as adding value. The customer does not care that that I have to make a trip to Home Depot. The customer does not care that this activity takes time.

In fact, it is annoying, as it delays the start and end of the project. These activities all delay the delivery of customer value.

So the project has some setup activities that are essential before the value-added work can start.

There may be others. There might be some planning. There may also be some estimation activity and some setting of expectations. The customer may be quoted a price for the job and a delivery date. (In this case, the customer for the project was my wife.)

When it comes to actually painting on the wood stain, it turns out that 42 sections of fence are too many to attempt in a single session. The velocity of painting was approximately four sections per hour. So the job was split up into six segments. If this were software development, we might have called these iterations or sprints. If it were manufacturing, we might call them batches. When I went to start a single segment of painting, I also had some setup activities. The first was to change clothes. Then I would set up the materials: I would move all the stain, brushes, and other tools from the garage to the location for that day's painting; only then could I start painting.

In summary, both projects and iterations have setup activities.

After a couple of hours of painting, I might want to take a break. Perhaps it is time for lunch. I can't just drop everything and start eating, though. First I must secure the stain by replacing the lid on the can, and then I must secure the brushes by either cleaning them or dropping them into a jar of water to prevent them from hardening while I take a break. Next I must personally clean up. I wash my hands and I change out of the work overalls; only then can I go and eat.

When the whole project is complete, I might have some extra wood stain remaining, and any full cans can be returned to Home Depot for a refund. So another trip is required.

It seems that both iterations and projects have a set of cleanup activities.

In economic terms, these setup and cleanup activities are referred to as transaction costs. Every value-added activity has associated transaction costs. These transaction cost activities are things that the customer may not see, most likely does not value, and to which they are ambivalent at best. The customer may be forced to pay the costs of these activities but would prefer not to. How often have you called a plumber to fix a washing machine or dishwasher and been asked for a $90 call out fee? This is a transaction cost. Would you prefer a lower fee? Would you choose a plumber who did not charge such a fee? The transaction costs do not add value. They may be necessary, but in Lean terms, they are waste.

So the first two types of waste are transaction costs, specifically: the setup, or front-end, and the cleanup, or back-end transaction costs.

If you consider this for software development activities, you will realize that all projects have a number of setup activities, such as project planning, resource planning and recruitment, budgeting, estimating, risk planning, communication planning, facilities acquisition, and so on. Most projects also have cleanup costs and other related back-end transaction costs, such as delivery to the customer, tear down of environments, retrospectives, reviews, audits, user training, and so on.

Iterations, too, have transaction costs, including the iteration planning and backlog selection (or requirements scoping), perhaps estimation, budgeting, resourcing, and environment setup. On the back end, they will have transaction costs that include integration, delivery, retrospective, and environment tear down.

Coordination Costs

Coordination is necessary as soon as two or more people try to achieve a common goal together. We invented language and communications systems in order to coordinate between human beings. When we agree to meet with friends, have drinks, dinner, and watch a movie on a Friday evening, we incur coordination costs. All the emails, text messages, and phone calls that are required to arrange a social evening are the coordination costs.

So coordination costs on projects are any activities that involve communicating and scheduling. When people on project teams complain that they can't do value-added work—such as analysis, development, or testing—because they are doing email, they are performing a set of coordination activities: Each email read and answered is a coordination activity. When they complain that they can't do value-added work—such as analysis, programming, or testing—because they are always in meetings, these meetings are also coordination activities.

In general, any form of meeting is a coordination activity, including favorites of the Agile community, such as daily standup meetings, unless the meeting is designed to produce a customer-valued deliverable. If three developers get together at a whiteboard and model a design for code they are about to implement, this is not a coordination activity, it is a value-added activity. Why? It is valued-added because it produces information that builds toward a complete customer-valued function.

If we view software and systems development as an information-arrival process, in which our starting point is no information, and complete information represents working functionality that meets the customer's needs and intent, then any information that arrives between the starting point and end point—that moves us forward toward working functionality that meets the customer's needs—is value-adding information.

If team members meet in order to create value-adding information, such as a design, a test, a piece of analysis, a section of code, then that meeting is not a coordination cost, it is a value-added activity.

However, if team members meet in order to discuss status, or task assignment, or scheduling that helps coordinate their actions and the flow of deliverables, that meeting *is* a coordination cost and should be regarded as wasteful. As such, you should seek to reduce or eliminate coordination meetings.

Hence, a 5-minute standup is better than a 15-minute standup if it achieves the same amount of coordination. A 15-minute standup is better than a 30-minute standup if it achieves the same amount of coordination.

You can think about reducing coordination activities by finding other, better ways of coordinating people.

One way is to empower team members to self-organize. Command-and-control type management, in which people meet in order to assign tasks to individuals in advance, is wasteful. It is better to let team members self-assign tasks. Self-organization generally reduces the coordination costs on a project. However, it requires information in order to work. Techniques within Kanban—such as the use of visual tracking of the value stream and visualization of work using a card wall and electronic tools and reports—provide coordination information that enables self-organization and reduces coordination costs on a project. Use of classes of service and visualizing them with colored cards or swim lanes on the card wall, together with the associated set of policies for the class of service, enable self-organization of scheduling and automatic prioritization. This is sometimes referred to as "self-expediting" (a term I associate with Eli Goldratt in reference to buffer management).

In general, the more information that can be made transparent to the knowledge workers on the team, the more self-organization will be possible and the fewer coordination activities will be required. Let transparency of work, process flow, and policies related to risk management displace coordination activities. Reduce waste through wider use of transparency.

How Do You Know if an Activity Is a Cost?

I have discovered that many people have trouble identifying wasteful activities. I have seen, for example, Agile advocates argue that daily standup meetings are value-added. I do not subscribe to this view. I cannot fathom that a customer cares whether a team holds standup meetings or not. What the customer wants is functionality that enables their goals, delivered in a timely manner with high quality. Whether or not a team needs to hold daily standup meetings to enable such delivery is neither here nor there from their perspective.

So how do you identify wasteful transaction costs or coordination activities?

I believe that you ask yourself, "If this activity is truly value-adding, would we do more of it?"

When you ask Scrum advocates who vehemently argue that daily standup meetings are value-added, whether they would hold the standup twice daily or whether they would lengthen it from 15 minutes to 30 minutes, they will surely reply, "No!"

"Well, if standup meetings are truly value-added," I reply, "then surely doing more of them would be a good thing!"

This is really the acid test that demonstrates the difference between a truly value-added activity and a transaction or coordination cost. Developing more customer requirements is clearly value-adding. You would do more of it if you could and the customer would gladly pay for it. Planning is clearly not value-added. The customer would not pay for more planning if he could avoid doing so.

So ask yourself, "Would we do more of this?" Challenge others with the same question about activities they undertake. If the answer is, "No," then consider how you might go about minimizing the time and energy spent on the activity, or how you might make the activity more effective, and hence, reduce the duration, frequency, or quantity of the activity.

It sometimes can be difficult to determine whether an activity is a transaction cost or a coordination cost. Some activities often look like both. I see this confusion all the time while teaching Kanban classes. As I do with class participants, I would urge you not to waste too much energy trying to determine the difference. What is important is that you have identified an activity as non-value-adding—and therefore wasteful—and you know that you want to reduce or eliminate that activity as part of a program of continuous improvement.

Failure Load

Failure load is demand generated by the customer that might have been avoided through higher quality delivered earlier. For example, a lot of help desk calls generate costs for a business. If the software or technology product or service were of higher quality, more usable, more intuitive, more fit-for-purpose, then there would be fewer calls. This would enable the business to reduce the number of call center personnel and reduce costs.

Lots of calls to a help desk tend to generate lots of production defect tickets. When selecting the functions in scope for a project or iteration, the business must choose between new ideas and production defects. Production defects aren't just software bugs; they include usability problems and other non-functional issues such as poor performance, lack of responsiveness under load or certain network conditions, and so on. The fix for a production defect against a non-functional requirement may appear like new functionality—a design for a new screen perhaps—but truly it isn't. It is failure load. That new screen design came about because of a usability defect in a previous release.

Failure load doesn't create new value; it enables value "left on the table" from a previous delivery. More than likely the earlier release of the product or service failed to deliver on its projected payoff function. Although some of this may be due to market variability or unpredictability, some of the shortfall will have been discovered to come from problems with the earlier release. Perhaps a bug in the product prevents the usage of some functionality. Because of this, potential customers are switched off the product and defer purchase or choose a rival product.

So the picture is muddy. Failure load still adds value. But what is important is that it adds value that should have been there already. Reducing failure load reduces opportunity cost of delay. Reduced costs mean more profits sooner. Reduced failure load means more of the available capacity can be spent on new functionality. Reduced failure load enables a business to pursue more market niches, with more product offerings. Reduced failure load enables more options. Reduced failure load may enable a reduction in team size, and hence, a reduction in direct costs.

Takeaways

❖ Waste can be classified into three main abstract categories: transaction costs, coordination costs, and failure load.

❖ The concept of waste is a metaphor.

❖ The waste metaphor does not work well for everyone, as waste is often necessary, though not specifically value-adding. As a result, I have replaced it with an economic cost model.

❖ To determine if an activity is truly wasteful ask, "Would we do more of this if we could?" If the answer is no, then the activity is some form of waste.

❖ Transaction costs come in two types: setup activities and cleanup, or delivery activities.

❖ Coordination costs are activities that are performed in order to assign people to tasks, schedule events, or coordinate the work of two or more people toward a common outcome.

❖ Failure load is new value-added work that is generated because of some earlier failing, such as a defect in the software, or a poor design or implementation that led to lack of customer adoption, a failure to realize a target payoff function, or a significant rise in help desk calls or service requests.

❖ Failure load uses capacity that could be used for new, innovative, additional customer-valued, and revenue-generating features.

❖ Turning ideas into working, customer-deliverable code more quickly maximizes potential value and minimizes waste.

❖ **Chapter 19** ❖

Sources of Variability

Variability in industrial processes has been studied since the early 1920s. The pioneer in this subject was Walter Shewhart. His techniques became the foundation for the quality assurance movement and are foundational elements of both the Toyota Production System and Six Sigma methods for quality and continuous improvement. Shewhart's techniques were adopted, developed, and advanced by W. Edwards Deming, Joseph Juran, and David Chambers. Their work was inspirational for Watts Humphrey and the founding members of the Software Engineering Institute at Carnegie Mellon University, who held the belief that the study of variation and its systematic reduction would bring great benefits to the software engineering profession.

There is a great deal of material published by Shewhart, Deming, and Juran on the study of variation and its use as a management technique and the foundation for a program of improvements. In addition, much has been published on the quantitative assessment method known as Statistical Process Control (SPC) that emerged as the main tool for studying variation and acting upon it. At the time of this writing, use of SPC is emerging with teams adopting Kanban. However, use of SPC is considered an advanced and high-maturity topic that will be addressed in a later book. Here, we will talk about variation in the most general terms and its simplest form.

Shewhart classified variability and variations in process performance into two categories: internal and external.

Internal sources of variation are variations that are under the control of the system in operation. With Kanban for software engineering and IT operations, we think of the system as a process that is defined by a set of policies that govern the system operation. They can be directly affected by changes made by individuals, the team, or management. Changes to policies change the system operation and its performance. Therefore, a change to the process definition represents a change that affects the internal sources of variation. Somewhat ironically, Shewhart named these internally generated variations "chance-cause variations." "Chance" implies that the variation is random and the randomness is a direct consequence of the system design. It does not imply that the randomness is evenly distributed or follows a standard distribution. Changes to the process design via changes in internal policies will affect any variation's mean, spread, and shape of distribution.

To use a general example, a batter in the game of baseball has a hit ratio (known as a batting average) that indicates how often he managed to hit a pitch that resulted in reaching first base or better. Different batters exhibit different ratios, with a typical range of about 0.100 to 0.350. On any given day, an individual batter may not achieve his typical hit ratio. This is determined by a number of factors such as pitcher selection, how well other players hit the ball, and the specific pitches thrown.

If we changed the rules of baseball to allow, say, four strikes before a batter is out, then we'd change the odds in favor of the batter versus the pitcher. The batter's average would increase as a result. Some better players might well achieve hit ratios in excess of 0.500 as a result of such a rule change. This is an example of modifying the system to modify the chance-cause variation within the system.

If we want to interpret this for a software development–specific example, an internal, chance-cause variation would be the number of bugs created per line of code, per requirement, per task, or per unit of time. The mean number, the spread, and the distribution of the bug (or defect) rate can be affected by changing the tools and process, such as by insisting on unit tests, continuous integration, and peer code reviews.

The process definition in use on your team, expressed as policies, represents the rules of the collaborative game of software development. The rules of the game determine the sources and quantity of internal variation. The irony lies in the notion that the "chance-cause" variations are actually directly under the control of the team and management through their ability to modify policies, change the process, and affect sources of internal variation.

External sources of variation are events that happen that are outside of the control of the immediate team or management. They are randomizations that come from other teams, suppliers, customers, and random "acts of God," as they are known in the insurance industry; for example, a two-week outage of a server farm caused by excessive flooding that resulted from unusually wet, stormy weather. External sources of varia-tion require a different approach to manage them. They cannot be directly affected by policies, but a process can be put in place to deal effectively with external variations. The body of knowledge that relates directly to this field is issue and risk management.

Shewhart named external variations "assignable-cause variations." By "assignable," he implied that someone (or a group of people) could easily point at the source of the problem and consistently describe it—such as, "There was a storm. It rained really hard and our server farm was flooded." Assignable-cause variations cannot be controlled by the local team or management but they can be predicted, and plans made and processes designed to cope with them gracefully.

Internal Sources of Variability

The software development and project-management process in place, coupled with the organizational maturity and capability of the individuals on the team determines the number of internal sources of variability and the degree of that variability.

To avoid confusion, Kanban must not be thought of as a software development lifecycle process or a project-management process. Kanban is a change-management technique that requires making alterations to an existing process: changes such as add-ing work-in-progress limits to it.

Work Item Size

The method of analysis used to break down requirements and itemize them for devel-opment has its own degree of variability. One dimension to this is the size of work items. Early literature describing the Extreme Programming method explained user stories as a narrative description of a feature as implemented and used by an end user, written on an index card. The only constraint was the size of the card. The effort required to create a user story was described as anything from a half-day to 5 weeks of work. Within a couple of years, a template for writing user stories had emerged from the community in London.

As a <user>, I want a <feature>, in order to <deliver some value>

The use of this template greatly standardized the writing of user stories. One of the creators of this approach, Tim McKinnon, reported to me in 2008 that he now had data to show that the average user story was 1.2 days of effort and the spread of variation was a half-day to about four days.

This is a specific example of reducing the chance-cause variation in the Extreme Programming method by asking the team to standardize user story writing around a given template. By doing so, Tim changed the rules of the game. The original rules asked team members to write stories on index cards in narrative form, and the new rules asked them to continue with index cards but to follow a specific sentence format. These changes are quite clearly under the influence and control of local managers. They are internal to the system. The size of a user story is controlled by chance-cause variation.

Work Item Type Mix

When all work is treated the same, and perhaps called by a single type, there is likely to be greater variation in size, effort, risk, or other factors. By breaking out work by specific type, it is possible to treat different types differently and to provide greater predictability.

For example, the Extreme Programming community developed type definitions for different sizes of stories. These acquired names like "epic" and "grain of sand." An epic might be a larger story that would take several people several weeks to develop, while a grain of sand might be a small story that could be completed by a single developer or pair of developers in a few hours. By adopting this nomenclature—Epic, Story, and Grain of Sand—we now have three types. For each individual type, the spread of variation will be lower than the spread would have been had all work been treated as one story.

Within a typical software development department there might be several types of work. There might be customer-valued new work with a name such as "feature," "story," or "use case." As just described, these might be stratified into size elements, or by some domain subtype or risk profile. There might be defect-removal work such as "production bug" or "(internal) bug." There might be maintenance work described as "refactoring," "re-architecting," or simply "upgrading." Software operating systems, databases, platforms, languages, APIs, and service architectures change over time and the code base needs to be updated to address the changes.

By using techniques to identify different work item types, we can change the mean and spread of variability and improve the predictability in the system for any one type of work.

An additional strategy to improve predictability is to allocate total WIP capacity by specific types. For example, with my maintenance team at Corbis, only two IT Maintenance items were permitted at any given time. This limited the capacity spent on API and database upgrades. This strategy is particularly useful when the types are divided out by size or effort required, such as epic, story, and grain of sand. By allocating specific capacity to each type, the responsiveness of the system is maintained and the predictability is greater.

Consider a team with a Kanban board on which there is a limit of two epics, eight regular stories, and four grains of sand. Two epics are in progress. A slot opens up in the queue for a regular story but there are none in the immediate backlog ready to start. The team has a choice of starting an epic or a grain of sand, or sticking to the type allocation and incurring some idle time.

If they start an epic, and a few days later a regular story shows up in the backlog, they would be unable to start the regular story for quite a while. This will increase the lead-time spread for regular stories.

Starting a smaller grain of sand is a better choice, as it might be finished before another regular story is ready to start. In that case, there is no impact, but there is a benefit from additional throughput. However, if they don't get lucky and they fail to complete the smaller item before a story is ready to start, then the lead-time spread for regular stories will be affected adversely, though not as badly as in the epic scenario.

Predictability and risk management typically should trump an opportunity to increase throughput, as business owners and senior managers value predictability more than throughput. Predictability builds and holds trust, a core Agile value, better than does delivering more with less reliability.

Class-of-Service Mix

If we consider the classes of service described in chapter 11, we can anticipate how variability might be affected by the mix of items. If an organization suffers from a lot of expedite requests, these will randomize everything else, increasing the lead-time mean and spread of variation, which reduces predictability of the whole system.

Expedite requests are essentially external variations and are described in the next section.

If the demand for other classes of service is fairly steady, the lead-time performance for each class should be fairly reliable. The mean and spread of variation should be measurable and should remain somewhat constant. This provides predictability. You can achieve this if the backlog is sufficiently large and full with a strong mix of each class. Allocate a WIP limit to each class of service. This will enable the mean and spread of variability for each class to settle down and the system will be predictable.

If the demand is variable—for example, if there are only a few fixed delivery date items and they tend to be seasonal—some action should be taken either to shape or control demand: Changes to the allocation of WIP limits across types should be instituted seasonally to anticipate demand changes, or, alternatively, changes to pull policies associated with the cascading set of classes of service should be made seasonally to cope with fluctuations in demand.

Consider a team with a WIP limit of 20, allocated as 4 fixed-date items, 10 standard class items, and 6 intangible class items. You can have a policy that these limits must be strictly adhered to, or you can loosen the rule and allow a standard or intangible item to fill a slot for a fixed date item when there is insufficient seasonal demand for fixed date items. These policies can be switched over at different times of year to improve the overall economic outcome and ensure that the system remains fairly predictable.

Irregular Flow

Irregular flow of work can be caused by both external and internal sources of variability. Every single item pulled through a kanban system will be different: different in nature to some degree, and different in size, complexity, risk profile, and effort required. The natural randomness of this will cause ebbs and surges in flow. A kanban system naturally copes with this as long as the WIP limits are enforced. However, greater variability from other sources, such as work item size, demand patterns, type mix, class of service mix, and external sources, require suitable buffering to absorb the ebbs and surges in flow through the system. Additional buffers may be required—and WIP limits will need to be larger—when there is more variability in the system. Greater WIP limits will result in longer lead times, but the smoother flow should reduce variability. Therefore, increasing a WIP limit to smooth flow will increase the mean lead time and reduce the range of lead-time variability. This is generally a more desirable outcome as managers, owners, and usually customers value predictability over the random chance of a shorter lead time or greater throughput.

Rework

Rework, whether from internal bugs being fixed before release or production defects displacing new customer-valued work, affects variability. If a defect rate is known, regularly measured, and fairly constant, the system can be designed to cope gracefully. Such a system will be economically inefficient, but it should be reliable. What causes a lack of predictability is when the defect rate is not anticipated correctly. Unplanned rework due to bugs lengthens lead times, tends to increase the spread of variation, and greatly reduces throughput. It seems to be very hard to plan for a specific defect rate, e.g., eight bugs per user story, let alone know or be able to predict their size and complexity. The best strategy for reduction of variability due to defects is to relentlessly pursue high quality with very low defect counts.

Making changes to the software development lifecycle process can greatly affect defect rates. Use of peer reviews, pair programming, unit tests, automated testing frameworks, continuous (or very frequent) integration, small batch sizes, cleanly de-fined architectures, and well-factored, loosely coupled, highly cohesive code design will greatly reduce defects. Changes that directly affect defect rates and indirectly improve the predictability of the system are directly under the control of the local management and the team.

External Sources of Variability

External sources of variability come from places that are not directly controlled by the software development process or project management method. Some of these will be from other parts of the business or the value stream, such as suppliers or customers. Other external sources include elements of the physical world that can't be easily antici-pated, predicted, or controlled—for example, a piece of equipment failing or adverse weather conditions.

Requirements Ambiguity

Poorly written requirements, ill-defined business plans, and lack of strategic planning, vision, or any other context-setting information may mean that a team member is unable to make a decision and therefore unable to complete a piece of work. A work item becomes blocked due to this inability to make a decision; new information is required to clarify the situation so that the team member can make a good-quality decision, allowing the work-in-progress to flow toward completion.

In order to reduce the impact of such blockages, the team and direct management need to implement an effective issue-management and resolution process, as described in chapter 20.

As a team and organization matures, it may be possible to discuss root-cause analy-sis and elimination. Blocking issues due to ambiguous requirements can be addressed by directly influencing the analysis processes used to develop requirements, and by improving the capability and skill level of those defining them. Measures such as these typically require the collaboration of other departments and managers and a will on the part of the business to improve.

At Corbis in 2007, this was achieved through a gradual process. First, the kanban system was implemented, including a visual board, an electronic tracking system, and the transparency that comes with that. The business became more and more involved and interested in the software development activity and in monitoring the process performance. A report was generated showing the number of open issues, the number of work items blocked, and the average time to resolve. (See Figure 12.6, the Issues and Blocked Work Items report, on page 144.)

When a requirement made it the whole way through to acceptance testing before it was rejected as not what the business really needed, the team reacted by creating a waste bin on the board and placing the ticket in it, as shown in Figure 19.1. Management then asked for a small set of electronic reports that showed work that had entered the system but had failed to make it the whole way through (Figure 19.2).

The combination of transparency, reporting, and building awareness of the impact and cost of poor requirements resulted in the business voluntarily changing its behavior. The waste report that showed the effect of poor requirements initially showed five to ten items per month. By the fifth month it was empty. The business had come to appreciate that by taking more care, they could avoid wasting capacity. They voluntarily collaborated to make the outcome of the system better. The net effect was root-cause elimination of the assignable-cause variations from poorly written requirements or ill-defined context information.

Figure 19.1. Board with waste bin

Although the software development team had taken actions to provide greater transparency and awareness, those actions did not directly affect the requirements-development process. The issue-management and resolution process merely mitigated the impact of blocking issues by raising awareness and reducing the time to resolve. The result was lower impact on the mean lead time and its spread of variation. The effects of transparency and reporting eventually resulted in an external change in process that eliminated the root cause of the problem.

This is anecdotal evidence that actions can be taken locally that will have an indirect effect on assignable-cause variations.

Expedite Requests

Expedite requests happen because of external events, such as an unexpected customer order, or due to some breakdown in a company's internal process, for example, a lack of communication that results in late discovery of some important requirement. Expedite requests are, by definition, assignable-cause variations, as the reason for the request is always known and therefore always "assignable."

Expediting is known in industrial engineering to be bad. It affects predictability of other requests. It increases mean lead time and the spread of variability and it reduces throughput. Evidence collected at Corbis throughout 2007 demonstrated that this industrial engineering result held true for software development processes: Expediting is undesirable even if it is being done to generate value.

The need for expediting can be reduced. Increasing slack capacity through improvements in throughput, automation, or increased resources will improve the ability to respond. Shorter lead times, greater transparency, and improving organizational maturity will reduce the need for expediting. Good teams adopting the Kanban approach have been shown to exhibit very little demand for expedite requests. In fact, at Corbis during 2007 there were only five such requests in total.

As with poor requirements, we can hope that transparency of process and good quality information regarding throughput, lead time, and due date performance will influence upstream behavior. We hope that demand will be shaped so that it is effectively understood early enough that it can be handled with a regular class of service rather than an expedite request.

One method of provoking this change is to agree on limiting the number of expedite requests that will be processed at any given time. At Corbis, this limit was one. By denying the business the ability to expedite anything they feel like, you force upstream people, such as sales or marketing people, to explore opportunities early and assess them effectively. If sales people are paid on commission and measured on revenue generated, failure to expedite something will hurt them. If it failed because

Rejected and Canceled Work Items

Report generated 4/27/2007 4:14:05 PM by CONTINUUM\DavidA; Last Warehouse update: 4/27/2007 3:31 PM

List Bugs, CRs and PDUs either Rejected or Canceled

ID	Work Item Type	Title	Dept	GTM Related	Business Priority	Submitted Date	Approved Date	Closed Date	Reason
2458	Bug	A lacinia leo justo vitae massa			1-Expedite	4/5/2007		4/5/2007	Overtaken by Events
2470	CR	Quisque vitae lacus sodales urna	Creative Services	Not Related to GTM		4/5/2007		4/5/2007	Overtaken by Events
1463	CR	Donec posuere malesuada sodales	Media Services	Not Related to GTM	2-High	11/26/2006		4/12/2007	Released
1443	CR	Pellentesque a. Duis et felis	Customer Experience	Not Related to GTM	2-High	11/26/2006		4/12/2007	Released
2703	CR	Semper turpis facilisis duis	HR	Not Related to GTM		4/26/2007		4/26/2007	Canceled

Figure 19.2   Rejected and Cancelled Work Report showing previous month's abandoned items

the WIP limit for expedite requests had been reached, then they will try harder in the future to gather enough information to post a request in time for it to be met with a regular class of service. Again, this is an example of some internal action that can be taken to indirectly affect an assignable cause of variation. A change to the system design that would normally affect the internal chance-cause variation has a secondary effect on the external assignable-cause variation.

Irregular Flow

Irregular flow of work can be caused by both chance- and assignable-cause variations as mentioned above. Assignable-cause variations that affect flow all result in blocked work. Problems such as ambiguous requirements, and environment– or specialist shared-resource availability are common reasons for assignable-cause blocked work.

Blocked work items require a strong discipline in and capability with issue management and resolution, as described in chapter 20. There are two approaches to dealing with the blocked work items issue.

The first approach will ease flow, but at the expense of lead time and possibly quality. You can improve flow by having a larger overall WIP limit—achieved through either explicit buffering or by using a policy with less restriction on WIP, for example, 3 things per person, on average, rather than 1.2 things per person. The greater WIP limit means that while something is blocked, the team members can be working on other items. I recommend this approach for immature organizations. The effects should be simple and lacking in drama. Lead times will be longer, but this may not be an issue in many domains. The spread of variation might be greater, and so lead times will be less predictable; however, they still may be more predictable through the use of a kanban system than they were before. The biggest drawback to using greater WIP limits is that there is less (or no) tension to provoke discussion and implementation of improvements. There is consequently no pressure to improve: The catalytic effect of kanban is lost.

The second approach is to pursue issue management and resolution relentlessly and, as the team matures, to move toward root-cause analysis and elimination with specific improvements designed to prevent assignable-cause variations in the future. In this approach, you leave the WIP limits, buffer sizes, and working policies fairly tight, and you cause the work to stop when things become blocked. Idle time for those people assigned to blocked work raises awareness of the blocking issue. It may cause a swarming behavior to try and fix the issue, which has been seen to encourage those idle team members to think about root causes and possible process changes that will reduce or eliminate the possibility of recurrence. Keeping WIP limits tight and pursuing issue management and resolution as a capability has been seen to create a culture of continuous improvement. I first saw this at Corbis in 2007, but there have been several other reports that emerged in 2009 at firms such Software Engineering Professionals in Indianapolis, IPC Media, and BBC Worldwide, both in London. There is now sufficient evidence to suggest that Kanban does provoke a culture that is focused on continuous improvement. The consistent process elements among the examples seem to be a willingness to enforce tight WIP policies, to mark work as blocked, to allow the line to stop, to incur idle time, and to pursue issue management and resolution as an organizational discipline. What results from this is a focus on root-cause analysis and elimination and the gradual introduction of improvements that both reduce assignable-cause variations and ignite a wider culture of continuous improvement.

Environment Availability

Environment availability is quite a typical assignable-cause issue that can have a significant effect on flow, throughput, and predictability. Environment outages often cause entire workflows to stall. A kanban system will bring visibility to the problem and its impact. The idle time incurred by enforcing a WIP limit has been seen to encourage collaboration on resolving the outage. When upstream folks such as developers and testers help systems-maintenance people to recover an environment, this behavior is often referred to as swarming. Swarming implies the concept that the team swarms together to work on a single problem until it is resolved. The nature of Kanban encourages teams to focus on lead time, throughput, and flow throughout the value stream. By aligning all the groups up and down the value stream with the same goals, there is an incentive for swarming behavior to emerge. Everyone wins when idle people volunteer to collaborate to resolve an issue that affects them even though it is not in their immediate work area or area of responsibility.

Other Market Factors

In October 2008, following the collapse of Lehman Brothers and series of related traumatic events in the financial sector, banks and investment firms in leading financial centers such as London and New York started to cancel or significantly modify IT projects in development. The reason was that their world had been turned upside down. They were fighting for their survival. Suddenly they needed to better understand their—and the market's—liquidity. It was no longer important to be delivering the latest exotic-commodity product. The market couldn't care less about investments. In the fall of 2008, financial enterprises were interested solely in solvency or insolvency, depending on how lucky they were.

This is a severe—but very real—example that shows how project portfolios and requirements for projects in progress can change dramatically. Reacting to these kinds of changes tends to distract teams and result in significant drops in throughput, dramatic increases in lead times, (often) drops in quality, and loss of predictability as the team recovers from the randomization that a fluctuation in the market causes to the internal workings of the project.

Clearly such events are assignable-cause variations. They need to be accommodated using risk-management strategies and tactics. There is a considerable body of knowledge on assignable-cause variation, or event-driven risk. Building a strong risk-management capability as part of an overall goal of improving organizational maturity will improve the predictability of a software engineering function whether it is using Kanban or not. However, kanban systems exhibit greater predictability when risk is managed well. This builds greater trust in the system.

Kanban systems have a number of other elements that assist in risk management. The WIP limit reduces risk, as only a small fraction of work is in progress at any given time. Allocation of WIP limits across work item types and classes of service help to manage risk and absorb assignable-cause variations. Other strategies are emerging, and it is likely that a subsequent book will also emerge, detailing advanced methods for improving Kanban and better risk-management tactics.

I've presented some material on managing risk with kanban systems that emerged from the use of Kanban at conferences during 2009, which is available online[23].

Difficulty Scheduling Coordination Activity

Another common source of assignable-cause variation that causes work to block and flow to become irregular is the challenge of coordinating external teams, stakeholders, and resources. One frequent reaction to coordination challenges is to schedule meetings with a regular cadence. In some instances this is very efficient. However, it won't always be possible.

Flow may be interrupted by a government or regulatory constraint that requires an audit or a signoff. The people required to perform this function may not be instantly available or may be difficult to schedule.

In the first instance, assignable-cause variation of this nature should be addressed by raising awareness and drawing attention to it with visibility and transparency. By marking items as blocked and raising visibility on to the source of the blockage, the management, team, and value-stream stakeholders will become aware of the impact of such coordination challenges.

This awareness should lead to some behavioral changes that improve the situation.

One tactic might be to examine the government and regulatory rules and decide whether everything needs to be assessed, approved, audited, or examined. Assuming that some risk profiling allows work to be bifurcated into two categories that do and do not need such a meeting to happen, either work item type or class of service can be used to separate out the work. You can then use allocation of the WIP limits to both types or classes to ensure smooth flow.

Takeaways

❖ The study of variation in industrial processes started in the 1920s with Walter Shewhart and evolved through the work of W. Edwards Deming, Joseph Duran, and David Chambers through the mid- to late twentieth century.

❖ A study of variation and its statistical analysis method is at the heart of both the Toyota Production System (and hence, Lean) and Six Sigma methods for process improvement.

❖ The work of W. Edwards Deming and Joseph Duran were a major inspiration for the work of the Software Engineering Institute at Carnegie Mellon University and the Capability Maturity Model (now Capability Maturity Model Integration, or CMMI).

❖ Shewhart divided sources of variation into two categories: those internal to the process or system and those external to the process or system.

❖ Internal variations are referred to as chance-cause variations.

❖ External variations are referred to as assignable-cause variations

❖ There can be many sources of chance-cause variations in the value stream of a software development lifecycle. Typical examples include work item size, type, class of service, irregular flow, and rework.

❖ There is a possibly endless list of sources of assignable-cause variations. Typical examples include requirements ambiguity, expedite requests, environment availability, irregular flow, market factors, staffing factors, and challenges in scheduling coordination or overhead activities.

❖ Chance-cause variation can be controlled through the use of policies that define the software development lifecycle and project-management processes in use.

❖ Assignable-cause variations can be managed through the use of issue-management and resolution capabilities as well as risk-management capabilities, and they can be reduced or eliminated through root-cause analysis and elimination capabilities.

❖ Kanban systems produce better economic outcomes when coupled to a solid, event-driven risk-management capability.

❖ Kanban also offers additional ways to manage risk such as allocation of a WIP limit across classes of service and work item types and the use of risk profiling to separate work into types or classes and process accordingly.

❖ Further work on advanced risk-management strategies and tactics with Kanban is ongoing, and it will be the topic of a future book.

❖ **Chapter 20** ❖

Issue Management and Escalation Policies

When work in our kanban system becomes blocked for any reason, it has become the convention to indicate this on the card wall by attaching a pink sticky note to the card, indicating the reason for the blockage. In electronic systems, there may be other ways to indicate that a work item is blocked, such as showing it with a red border. Preferably, the features of the electronic system should allow the reason for the blockage to be tracked separately, or blocking issues should be tracked as first-class work items linked to the customer-valued item that is dependent upon resolution of the issue.

During the writing of this book, I have noticed that some people attempting Kanban for the first time refer to these blocked items as bottlenecks. This is wrong. A blocked item might be clogging the pipe and restricting flow, but it is not a bottleneck, such as it is described in chapter 17: It is neither a capacity-constrained resource nor a non-instant availability resource. In the same manner, a cork in a bottle-neck is not a bottleneck. If you want to restore flow from the bottle, you simply remove the cork.

It's dangerous to think of blocked work items as bottlenecks be-cause it leads to the wrong kind of thinking toward resolving the problem. Blocked work items should be treated as special cause varia-tions rather than as bottlenecks. What is similar is the desired out-come. In both cases—a bottleneck or a blocked work item—we want to resolve the issue in order to improve flow.

Blocked work items require an organization to develop a capability for issue management and resolution to restore flow as quickly as possible, as well as a capability at root-cause analysis and resolution to prevent recurrence. The latter capability was discussed in chapter 19 as removal of special cause variations. The former is discussed here.

Managing Issues

It isn't good enough simply to mark and track work as blocked. Many early Agile software development tools allowed only for this capability. While it is useful information to know that an item, a story, a feature, or a use case is blocked, my observation of teams around the world has been that knowing something is blocked does not lead to developing a strong capability for getting it unblocked.

It is essential to track the reason for the blockage and to treat it as a first-class work item, albeit a failure-load work item. A special work item type, Issue, is set aside for this purpose. Issues are tracked using pink tickets (Figure 20.1). They should be assigned a tracking number, and a team member—usually the project manager—should be assigned to ensure resolution.

When a team member working on a customer-valued item is unable to proceed, he or she should mark the item as blocked, attach a pink ticket to it describing the reason for the blockage, and create an Issue work item in the electronic tracking tool. The issue should be linked to its original work item (see Figure 20.1). Some examples are: ambiguity in the requirements, and a knowledgeable person is not available to instantly resolve the ambiguity; an environment setup is required and an engineer to perform this task is currently unavailable; or a specialist is required to work on the item and that person is unavailable due to vacation, sickness, or other out-of-office time.

As discussed in chapter 7, maintaining flow should be the main focus of discussion at the daily standup meeting. Hence, the meeting should focus on discussing blockages and progress toward the resolution of the issues. The meeting should focus heavily on the pink tickets. Questions should be asked about who is working on resolving the issue and the status of progress on resolution. Does the issue need to be escalated? If so, to whom?

Idle team members should be encouraged to volunteer to track down issues and generally to swarm on problems and assist however they can to resolve them and restore flow to the system. A team with strong self-organization capability will tend to do this naturally. Team members will volunteer to help resolve issues. Where that self-organization capability has yet to emerge, the project manager may need to assign team members to work issues to resolution.

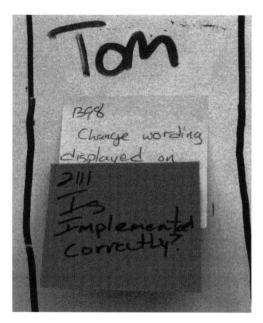

Figure 20.1 Pink blocking issue (or impediment) item attached to the Change Request work item directly affected

Like all other work items, issues should be tracked. The tracking should include a start- and an end date and a link to all affected customer-valued work items. Note that a single issue may be blocking more than one customer-valued item. This is another good reason for tracking issues as independent work items and for having Issue as a distinct work item type. When choosing an electronic tool for tracking your kanban system, be sure to pick one that supports issue tracking as a first-class type or a tool that is sufficiently customizable that you can create a type of work called Issue and designate that it will be displayed using pink (or red) cards.

Escalating Issues

When the team is unable to resolve an issue on its own, or an external party is required to resolve an issue and is unavailable or unresponsive, the issue must be escalated to a more senior manager or other department.

It is important for the organization to develop a strong capability at issue escalation. Without it, maintaining and restoring flow after a blockage can be problematic.

The foundation of a good escalation capability is a documented escalation policy or process. Chapter 15 discusses the power of developing organizational policies in a collaborative fashion. Escalation policies should be developed in such a collaborative fashion and a consensus agreed upon among departments involved in the value stream. The escalation policies should be widely known and understood, and a document (or web site) describing them should be readily available to all team members. There should be no ambiguity over how and where to escalate a problem. By taking the time to define escalation paths and write policy around it, the team knows where to send issues for resolution. This saves time figuring out to whom an issue should be escalated and it sets expectations for those more senior individuals that they are expected to be a part of the process. Senior managers need to take responsibility to resolve issues. This will help to maintain flow and ultimately to minimize cost of delay (or optimize payoff from speedy delivery.)

Tracking and Reporting Issues

As stated earlier, issues should be tracked as first-class work items with their own work item type. The convention has evolved to use pink or red cards or sticky notes to visualize issues (Figure 20.2). A start date, an end date, an assigned team member, a description of the issue, and links to the blocked customer-valued items are the minimum requirements for an issue tracking system (Figure 20.3). Some history of efforts made at resolution, a history of assigned individuals, an indication of the escalation

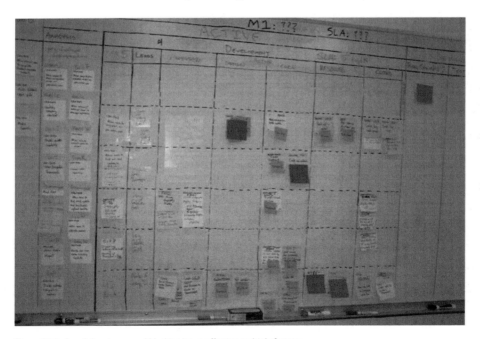

Figure 20.2 Board showing several blocking issues affecting multiple features

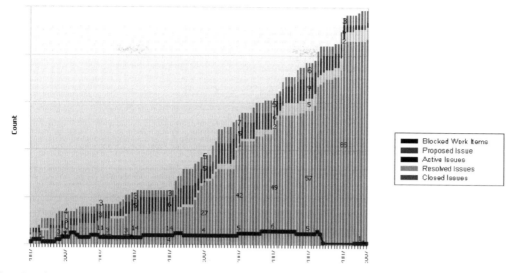

Figure 20.3 Issues Cumulative Flow Diagram(CFD) with overlaid Blocked Work Items graph

path, an estimated time to resolution, an impact assessment, and suggested root-cause fixes for future prevention may also be useful fields to track.

Even though pink tickets on the card wall provide a strong visualization of how many items are currently blocked, it is also useful to track and report issues in other ways. A cumulative-flow diagram of issues and blocked work items provides a strong visual indicator of the organizational capability at issue management and resolution. The trend in blocked work items over time indicates whether a capability of root-cause analysis and resolution—improvement opportunities to eliminate assignable-cause variations—is developing. A tabular report of current issues, assigned individuals, status, anticipated resolution date, affected work items, and potential impact may also prove useful for day-to-day management on larger projects.

These reports should be presented at each operations review and time should be set aside to discuss the emergence and maturity of organizational capability of issue management and resolution and root-cause analysis and resolution. The organization should be aware of the failure-load impact of blocking issues. This will enable objective decisions about improvement opportunities and the likely benefits of investment in root-cause fixes to prevent special-cause variations.

Takeaways

❖ Kanban systems should have a first-class work item type, Issue, used to track problems causing other customer-valued work to block.

❖ It has become the convention to use pink (or red) sticky notes on a card wall to visualize blocking issues.

❖ Pink issue tickets are attached to the items that are blocked.

❖ A strong capability for issue management and resolution is essential to maintain flow.

❖ Blocked work items and issues are not bottlenecks. They should be managed as special-cause variations rather than as capacity-constrained resources or non-instant availability resources.

❖ Issue management should be a strong focus of daily standup meetings.

❖ A strong capability for issue escalation is essential as part of a strong capability for issue management.

❖ Escalation policies should be clearly defined and documented and all team members should be aware of them.

❖ Escalation policies work better when they are agreed upon collaboratively by all departments involved in the value-stream.

❖ Issues should be tracked electronically.

❖ Some reporting based on electronic data will facilitate day-to-day issue management and resolution on larger projects.

❖ Use an Issues and Blocked Work Items cumulative-flow diagram to visualize the development of capability for issue management and resolution and root-cause analysis and resolution.

❖ Endnotes ❖

1. Anderson, David J. *Agile Management for Software Engineering: Applying the Theory of Constraints for Business Results.* Upper Saddle River, NJ: Prentice Hall, 2003.

2. Beck, Kent. *Extreme Programming Explained: Embrace Change.* Boston: Addison Wesley, 2000.

3. Beck, Kent et al., "The Principles Behind the Agile Manifesto." http://www.agilemanifesto.org/principles.html.

4. Goldratt, Eliyahu M. *What is this thing called The Theory of Constraints and How should it be implemented?* Great Barrington, MA: North River Press, 1999.

5. Anderson, David J., and Dragos Dumitriu, "From Worst to Best in 9 Months: Implementing a Drum-Buffer-Rope Solution in Microsoft's IT Department," Proceedings of the TOCICO World Conference, Barcelona, November 2005.

6. Belshee, Arlo. "Naked Planning, Promiscuous Pairing and Other Unmentionables." 2008 Agile Conference, podcast. http://agiletoolkit.libsyn.com/index.php?post_id=400364.

7. Hiranabe, Kenji. "Visualizing Agile Projects Using Kanban Boards." *InfoQ,* August 27, 2007. http://www.infoq.com/articles/agile-kanban-boards.

8. Hiranabe, Kenji, "Kanban Applied to Software Development: From Agile to Lean," *InfoQ*, January 14, 2008. http://www.infoq.com/articles/hiranabe-lean-agile-kanban.

9. Augustine, Sanjiv. *Managing Agile Projects.* Upper Saddle River, NJ: Prentice Hall, 2005.

10. Highsmith, Jim. *Agile Software Development Ecosystems.* Boston: Addison Wesley, 2002.

11. The Nokia Test is attributed in origin to Bas Vodde, described here by Jeff Sutherland, who has adopted and updated it. http://jeffsutherland.com/scrum/2008/08/nokia-test-where-did-it-come-from.html.

12. Beck et al., "The Principles Behind the Manifesto." http://www.agilemanifesto.org/principles.html.

13. Jones, Capers. *Software Assessment Benchmarks and Best Practices.* Boston: Addison Wesley, 2000.

14. Ambler, Scott. *Agile Modeling: Effective Practices for Extreme Programming and the Unified Process.* Hoboken, N.J.: Wiley, 2002.

15. Chrissis, Mary Beth, Mike Konrad, and Sandy Shrum. *CMMI: Guildelines for Process Integration and Product Improvement,* 2d ed. Boston: Addison Wesley, 2006.

16. Sutherland, Jeff, Carsten Ruseng Jakobsen, and Kent Johnson. "Scrum and CMMI Level 5: A Magic Potion for Code Warriors." Proceedings of the Agile Conference, Agile Alliance/IEEE, 2007.

    Jakobsen, Carsten Ruseng and Jeff Sutherland. "Mature Scrum at Systematic." *Methods & Tools,* Fall 2009. http://www.methodsandtools.com/archive/archive.php?id=95.

17. Larman, Craig and Bas Vodde. *Scaling Lean & Agile Development: Thinking and Organizational Tools for Large-Scale Scrum.* Boston: Addison Wesley, 2008.

18. Willeke, Eric, with David J. Anderson and Eric Landes (editors) *Proceedings of the Lean & Kanban 2009 Conference.* Bloomington, IN: Wordclay, 2009.

19. Beck et al, Principles Behind the Agile Manifesto, 2001, http://www. agilemanifesto.org/principles.html

20. Anderson, David J. "New Approaches to Risk Management." Agile 2009, Chicago, Illinois. http://www.agilemanagement.net/Articles/Papers/Agile2009-NewApproachesto.html.

❖ Acknowledgments ❖

Every book published represents a significant project-management and coordination effort and involves a whole team of people, of whom the author is really only a small part. This book would not have happened without the invaluable efforts, hard work, and dedication of Janice Linden-Reed and Vicki Rowland. I would like to thank them for their incredible patience and perseverance to get this manuscript to press against a tight deadline (with a high cost of delay.)

I'd like to thank Donald Reinertsen for prompting me to try Kanban and for giving me an early forum to talk about it publicly. I'd also like to thank him for his kind words in the Foreword and his efforts to build a long-lived and thriving community through the formation of the Lean Software & Systems Consortium.

I'd like to thank Karl Scotland, Joe Arnold, and Aaron Sanders, together with Eric Willeke, Chris Shinkle, Olav Maassen, Chris Matts, and Rob Hathaway. Their early enthusiasm and adoption of Kanban led directly to the formation of the now-thriving community and the viral spread of the method around the globe. Without their support there would be no demand for this manuscript, and Kanban would be an obscure method used at a couple of companies in the Pacific Northwest of the United States, rather than the exciting new approach being used by teams on every continent—from five-person startups in Cambodia to 300-year-old insurance companies in the Netherlands to large oil companies in Brazil and outsource vendors in Argentina, as well as media companies in London, Los Angeles, and New York and many, many more throughout the world. Kanban adoption is a phenomenon, and it wouldn't have happened without the fortuitous meeting of minds that occurred in August 2007 in Washington, D.C., at the Agile 2007 conference.

This book would not be such a useful tool and enjoyable reading experience were it not for the thoughtful comments and constructive feedback from a large group of manuscript reviewers. I'd like to call particular attention to the contributions of Daniel Vacanti, Greg Brougham, Christina Skaskiw, Chris Matts, Bruce Mount, Norbert Winklareth, and, again, Janice Linden-Reed. Each of them produced a strategic, thoughtful review of one or more early versions of the manuscript that led to significant restructuring of the content. The result is a much better book that is easier to read, easier to understand, and that will be a more useful long-term tool for the community.

In addition, there were many more members of the community who contributed feedback and edits that were all considered carefully as the manuscript emerged and evolved during 2009 and 2010. Thanks to: Jim Benson, Matthias Bohlen, Joshua Kerievsky, Chris Simmons, Dennis Stevens, Arne Roock, Mattias Skarin, Bill Barnett, Olav Maassen, Steve Freeman, Derick Bailey, John Heintz, Lilian Nijboer, Si Alhir, Siddharta Govindaraj, Russell Healy, Benjamin Mitchell, David Joyce, Tim Uttormark, Allan Kelly, Eric Willeke, Alan Shalloway, Alisson Vale, Maxwell Keeler, Guilherme Amorim, Reni Elisabeth Pihl Friis, Nis Holst, Karl Scotland, and Robert Hathaway.

I'd like to thank my tireless office manager, Mikiko Fujisaki, who keeps the wheels turning at David J. Anderson & Associates, Incorporated, and without whom I'd never have found the time to write this book.

My old friend and colleague Pujan Roka kindly offered to draw the cover illustration. Pujan is a talented comic artist, and is also a published author in his own right, with two significant management books to date. Learn more about him and his publications at http://www.pujanroka.com/

The community has been so generous in its adoption and enthusiasm for Kanban, which has extended to the kind offers to translate this manuscript into local languages. I'd like to thank Jan Piccard de Muller, Andrea Pinto, Eduardo Bobsin, Arne Roock, Masa Maeda, and Hiroki Kondo, who are already hard at work producing the French, Portuguese (Brazilian), German, Spanish, and Japanese translations. I am sure that their efforts will help to spread Kanban adoption around the world and expand the community and the enthusiasm for the method in their respective regions.

I'd like to acknowledge Nicole Kohari, Chris Hefley, David Joyce, Thomas Blomseth, Jeff Patton, and Steve Reid, all of whom contributed images for use in the book.

And finally, I'd like to thank my good friend Dragos Dumitriu, now at Avanade, and my team at Corbis: Darren Davis, Larry Cohen, Mark Grotte, Dominica Degrandis, Troy Magennis, Stuart Corcoran, together with Rick Garber, Corey Ladas, and Diana Kolomiyets. Without all of them, Kanban would never have happened. Their efforts at implementing and using it created the examples and stories from which we've all learned and subsequently adopted and adapted solutions for new and ever more challenging situations. Without them there would be no book, no community, and no growing band of delighted customers who are enjoying high-quality software developed regularly and quickly, when and where they need it, with agility, in response to the natural demand of their industry and user base.

Our Kanban journey continues and hopefully with this book, you have been persuaded to become part of it.

David J. Anderson
On the Kanban evangelization trail,
somewhere in Europe, April 2010

❖ Index ❖

❖ **About the Author** ❖

David J. Anderson leads a management consulting firm focused on improving performance of technology companies. He has been in software development nearly 30 years and has managed teams on agile software development projects at Sprint, Motorola, Microsoft, and Corbis. David is credited with the first implementation of a kanban process for software development, in 2005. David was a founder of the Agile movement through his involvement in the creation of Feature Driven Development. He was also a founder of the Agile Project Leadership Network (APLN), a founding signatory of the Declaration of Interdependence, and a founding member of the Lean Software and Systems Consortium. He moderates several online communities for lean/agile development. He is the author of *Agile Management for Software Engineering: Applying the Theory of Constraints for Business Results*. Most recently, David has been focused on creating a synergy of the CMMI model for organizational maturity with Agile and Lean methods through projects with Microsoft and the SEI. He is a co-author of the SEI's Technical Note "CMMI and Agile: Why not Embrace Both!" He is based in Sequim, Washington, USA.

Additional Kanban Resources

David J. Anderson and Associates
http://djandersonassociates.com

The Limited WIP Society
http://www.limitedwipsociety.org

Kanban Development Yahoo! Group
http://finance.groups.yahoo.com/group/kanbandev/

Kanban101
http://www.kanban101.com

Made in the USA
San Bernardino, CA
26 September 2014